T0331143

Introduction to Traveling Waves

Introduction to Traveling Waves is an invitation to research focused on traveling waves for undergraduate and masters level students. Traveling waves are not typically covered in the undergraduate curriculum, and topics related to traveling waves are usually only covered in research papers, except for a few texts designed for students. This book includes techniques that are not covered in those texts.

Through their experience involving undergraduate and graduate students in a research topic related to traveling waves, the authors found that the main difficulty is to provide reading materials that contain the background information sufficient to start a research project without an expectation of an extensive list of prerequisites beyond regular undergraduate coursework. This book meets that need and serves as an entry point into research topics about the existence and stability of traveling waves.

Features

- Self-contained, step-by-step introduction to nonlinear waves written assuming minimal prerequisites, such as an undergraduate course on linear algebra and differential equations.
- Suitable as a textbook for a special topics course, or as supplementary reading for courses on modeling.
- Contains numerous examples to support the theoretical material.
- Supplementary MATLAB codes available via GitHub.

Anna R. Ghazaryan is a Professor of Mathematics at Miami University, Oxford, OH. She received her Ph.D. in 2005 from the Ohio State University, Ohio. She is an applied analyst with research interests in applied dynamical systems, more precisely, traveling waves and their stability.

Stéphane Lafortune is Professor of Mathematics at the College of Charleston in South Carolina. He earned a dual PhD in Physics from the Université de Montréal, Canada and Université Paris VII, France in 2000. He is an applied mathematician who works on nonlinear waves phenomena. More precisely, he is interested in the theory of integrable systems and in the problems of existence and stability of solutions to nonlinear partial differential equations.

Vahagn Manukian is an Associate Professor of Mathematics at Miami University, Ohio. He obtained a M.A. Degree Mathematics from SUNY at Buffalo and a Ph.D. in mathematics from the Ohio State University in 2005. Vahagn Manukian uses dynamical systems methods such as local and global bifurcation theory to analyze singularly perturbed nonlinear reaction diffusions systems that model natural phenomena.

Introduction to Traveling Waves

Anna R. Ghazaryan
Miami University, USA

Stéphane Lafortune
College of Charleston, USA

Vahagn Manukian
Miami University, USA

CRC Press
Taylor & Francis Group
Boca Raton London New York

CRC Press is an imprint of the
Taylor & Francis Group, an **informa** business
A CHAPMAN & HALL BOOK

First edition published 2023

by CRC Press
6000 Broken Sound Parkway NW, Suite 300, Boca Raton, FL 33487-2742

and by CRC Press
4 Park Square, Milton Park, Abingdon, Oxon, OX14 4RN

Library of Congress Cataloging-in-Publication Data
Names: Ghazaryan, Anna R., author.	Lafortune, Stéphane (Mathematician), author.	Manukian, Vahagn, author.
Title: Introduction to traveling waves / Anna R. Ghazaryan (Miami University, USA), Stéphane Lafortune (College of Charleston, USA), Vahagn Manukian (Miami University, USA).
Description: First edition.	Boca Raton : C&H/CRC Press, 2023.	Includes bibliographical references and index.
Identifiers: LCCN 2022022202 (print)	LCCN 2022022203 (ebook)	ISBN 9780367707057 (hardback)	ISBN 9780367707026 (paperback)	ISBN 9781003147619 (ebook)
Subjects: LCSH: Nonlinear waves.	Reaction-diffusion equations.
Classification: LCC QA927 .G439 2023 (print)	LCC QA927 (ebook)	DDC 531/.1133--dc23/eng20220910
LC record available at https://lccn.loc.gov/2022022202
LC ebook record available at https://lccn.loc.gov/2022022203

ISBN: 978-0-367-70705-7 (hbk)
ISBN: 978-0-367-70702-6 (pbk)
ISBN: 978-1-003-14761-9 (ebk)

DOI: 10.1201/9781003147619

Typeset in Latin Modern font
by KnowledgeWorks Global Ltd.

Publisher's note: This book has been prepared from camera-ready copy provided by the authors.

To our families.

Contents

Preface   xi

CHAPTER   1 ▪ Nonlinear traveling waves   1

1.1   TRAVELING WAVES   1

1.2   REACTION-DIFFUSION EQUATIONS   5

1.3   TRAVELING WAVES AS SOLUTIONS OF REACTION-DIFFUSION EQUATIONS   8

1.4   PLANAR WAVES   12

1.5   EXAMPLES OF REACTION-DIFFUSION EQUATIONS   14

    1.5.1   Fisher-KPP equation   14

    1.5.2   Nagumo equation   18

1.6   OTHER PARTIAL DIFFERENTIAL EQUATIONS THAT SUPPORT WAVES   21

    1.6.1   Nonlinear diffusion, convection, and higher order derivatives   21

    1.6.2   Burgers equation   21

    1.6.3   Korteweg-de Vries (KdV) equation   22

CHAPTER   2 ▪ Systems of Reaction-Diffusion Equations posed on infinite domains   27

2.1   SYSTEMS OF REACTION-DIFFUSION EQUATIONS   27

2.2   EXAMPLES OF REACTION-DIFFUSION SYSTEMS   31

    2.2.1   FitzHugh-Nagumo system   31

    2.2.2   Population models   34

2.2.3	Belousov-Zhabotinski reaction	38
2.2.4	Spread of infection disease	41
2.2.5	The high Lewis number combustion model	42

CHAPTER   3 ■ Existence of fronts, pulses, and wavetrains   49

3.1	TRAVELING WAVES AS ORBITS IN THE ASSOCIATED DYNAMICAL SYSTEMS	49
3.2	DYNAMICAL SYSTEMS APPROACH: EQUILIBRIUM POINTS	53
3.3	EXISTENCE OF FRONTS IN FISHER-KPP EQUATION: TRAPPING REGION TECHNIQUE	58
3.3.1	Existence of fronts in Nagumo equation	64
3.3.2	Rotated vector fields and existence of a heteroclinic orbit between A and C for some $c \neq 0$	69
3.4	EXISTENCE OF FRONTS IN SOLID FUEL COMBUSTION MODEL	71
3.5	WAVETRAINS	75

CHAPTER   4 ■ Stability of fronts and pulses   83

4.1	STABILITY: INTRODUCTION	83
4.2	A HEURISTIC PRESENTATION OF SPECTRAL STABILITY FOR FRONT AND PULSE TRAVELING WAVE SOLUTIONS	86
4.2.1	Eigenvalue problem	87
4.2.2	Spectrum and spectral stability	97
4.3	LOCATION OF THE POINT SPECTRUM	109
4.3.1	Spectral Energy Estimates	109
4.3.2	Evans function	119
4.3.2.1	Definition of the Evans function	119
4.3.2.2	Gap Lemma	125
4.3.2.3	Evans function computation: scalar equations	126

4.3.2.4   Evans function computation: systems
of equations   134

4.4   BEYOND SPECTRAL STABILITY   144

Bibliography   149

Index   159

Preface

Purpose and the audience

We designed this text as *an invitation to research focused on traveling waves* for undergraduate and master level students. Traveling waves are not covered in the undergraduate curriculum. Topics related to traveling waves are usually covered in research papers, with the exception of a few texts designed for students, but these texts do not teach the techniques and methods presented in this book. Through our experience, we know that when involving undergraduate and graduate students in a research topic related to traveling waves, the main difficulty is to provide a student with reading materials that contains a background information sufficient to start a research project without an expectation of an extensive list of prerequisites beyond regular undergraduate coursework. This book should fill this void and serve as an entry point into research topics about the existence and stability of traveling waves.

The book is self-contained and includes numerous examples. We believe that it is suitable for a text for independent studies or a special topics course.

The book starts from the basics and then explains the techniques and concepts related to traveling waves assuming no more than an undergraduate level knowledge in single and multivariable Calculus, Linear Algebra, and Differential Equations. We assumed that the students are familiar with the concepts of the phase space and equilibrium solutions. The explanations of the discussed techniques are step-by-step. We tried not to overwhelm the reader with theoretical concepts but rather show some techniques that are applicable to a variety of equations. Most of the techniques are demonstrated by working on examples and we tried to include enough detail for the students to feel comfortable to try their hand on different examples of equations or systems of equations. We made it clear in the book when we avoided going into depth in one or another topic, being careful not to create an overly simplistic look at this research topic.

The book touches on both theoretical and numerical aspects of the analysis of nonlinear waves. The numerical aspects are done through some MATLAB coding and the specialized software package STABLAB. We made the codes that we use in the book available on GitHub. We assume only some basic familiarity with MATLAB, comparable with, for example, MATLAB Onramp, which is a free, two-hour introductory tutorial available at MathWorks.com.

Our hope is that this book will help to provide a positive early research experience and, possibly, work as a gateway to further research at higher, graduate levels.

MATLAB is a registered trademark of The Mathworks, Inc. The name MATLAB is used in this book without the inclusion of the trademark symbol, in an editorial content only; no infringement of trademark is intended.

STABLAB: A MATLAB-Based Numerical Library for Evans Function Computation (June 2015)) is an open-source, free MATLAB-based interactive toolbox that is used for stability analysis of traveling waves such as fronts and pulses. Developed by B. Barker, J. Humpherys, J. Lytle, and K. Zumbrun, it is available in the Github repository, nonlinear-waves/stablab [3,4].

Acknowledgments

We would like to extend our thanks to Blake Barker for his continuous support with STABLAB. Blake was always willing and available to answer questions and offer help with that software that we used for the computation of the Evans function. He also graciously read the part of the book dealing with the description of STABLAB and provided feedback.

We also would like to thank Alessandra Manukian-Kazarian. While working on her Bachelor of Arts in Interactive Media Studies degree at Miami University, she produced all of the figures in this book that are not outputs of MATLAB or STABLAB and assisted with the design of the graphics for the cover. We appreciate her infinite patience, enthusiasm, and continuous support.

MATLAB® is the registered trademark of The MathWorks, Inc. For
product information, please contact:
The MathWorks, Inc.
3 Apple Hill Drive
Natick, MA, 01760-2098 USA
Tel: 508-647-7000
Fax: 508-647-7001
E-mail: info@mathworks.com
Web: https://www.mathworks.com

CHAPTER 1

Nonlinear traveling waves

1.1 TRAVELING WAVES

Mathematical models for many naturally occurring or experimentally built systems include nonlinear partial differential equations posed on spatially unbounded domains. This is justified when the internal wavelengths of these systems are much smaller than the sizes of the underlying physical domains, for example, the distance between two sand ripples compared to the length of the beach, or when the size of a spatially localized structure is much smaller than the underlying domain, for example, the center of the tornado compared to the tornado itself.

To understand the information provided by the partial differential equations, one usually seeks its equilibrium solutions. Assume that the independent variables in the equation are $x \in \mathbb{R}$, which represents the space, and $t \geq 0$, which represents the time. The equilibrium solutions are time-independent, so they are also called stationary. These may be spatially homogeneous or constant solutions or patterns, which depend on the spatial variable but not on the time variable.

A conceptual example of constant equilibrium solutions are the two equilibria of a simple pendulum system that consists of a rod with a hinge on one end and a mass attached to the opposite end. If gravity is the only force acting on the mass, the position where the mass is exactly under the hinge is an equilibrium position. If the mass happens to be in this position and is not perturbed, it will stay there forever. Another equilibrium position is when the mass is exactly above the hinge. The mass will stay there as long as there are no perturbations of any

DOI: 10.1201/9781003147619-1

kind, which is not a very realistic assumption. This is actually a very convenient example to introduce an intuitional concept of the stability of an equilibrium. An equilibrium is stable if small perturbations are not enough to disturb the system, like in the first equilibrium for the pendulum. If the mass is moved away slightly from this equilibrium position, the pendulum after oscillating for a while will converge to the equilibrium. On the other hand, the equilibrium directly above the hinge is unstable. Even a slightest perturbation causes the system to seek the other equilibrium.

Another example of a constant state is related to combustion theory where burning of the fuel is modeled by partial differential equations. The physical state where the fuel is completely gone is a constant equilibrium state.

Patterns on animal skins are examples of patterns described by non-linear partial differential equations. They are "usually" not moving and come in the form of stripes and spots, or rings, or combinations of those. Vegetation growth in semi-arid areas sometimes occurs in well-defined patterns as well. The health of the shellfish populations like mussels on mussel beds may be evaluated by the presence of patterns [100]. There exist examples of "moving patterns". For example, mutant mice were "designed" in laboratory conditions, so their skin color appeared to oscillate [94]. The latter case falls under the umbrella of traveling waves that we discuss below.

The dynamics on unbounded domains is essentially different from that on bounded domains. In particular, partial differential equations posed on unbounded domains may support special solutions which are called traveling waves. Traveling waves are solutions of the underlying partial differential equation that preserve their shape while propagating with a certain velocity. Traveling waves are not stationary in the original spatial variable. These are solutions of the partial differential equation that depend on a specific combination of the variables x and t, more precisely, on the combination $x - ct$, where c is the constant that represents the velocity with which the solution moves. So, in the moving coordinate $\xi = x - ct$, the solution is stationary since the profile of the wave does not change in time. The partial derivative of the wave profile with respect to the time variable is zero, and therefore the wave may be sought as a solution of the ordinary differential equation, with the derivatives taken with respect to the moving variable ξ. These ordinary differential equations are called traveling wave equations.

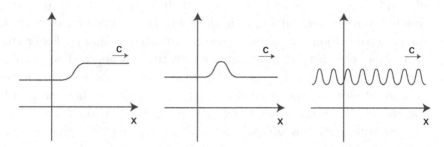

Figure 1.1 From the left to the right: a front, a pulse, and a periodic wavetrain in a stationary coordinate frame.

Traveling waves are translationally invariant in the sense that if a function $u_0(\xi)$ solves the traveling wave equation then $u_0(\xi + \xi_0)$ is also a solution for any constant ξ_0. Since a shift of a traveling wave is a traveling wave as well, traveling waves always exist in families and never alone.

The main types of traveling waves are fronts, pulses, and periodic wavetrains. Some traveling waves have constant (independent of the spatial variable) equilibria as limits at $+\infty$ and $-\infty$. These equilibria are called rest states of the associated wave. If a traveling wave asymptotically connects two distinct spatially homogeneous equilibria, it is called a front. If the rest states of a wave are the same, the wave is called a pulse. Periodic wavetrains do not have asymptotic limits at $+\infty$ and $-\infty$. These are periodic solutions of the traveling wave equations. Figure 1.1 contains a schematic illustration of fronts, pulses, and wavetrains. Traveling waves in partial differential equations and systems may exist for a specific discrete value of the speed c or for a continuum of values of the speed c.

In applications, a full investigation of a traveling wave includes a discussion of its stability. A stable wave is a wave that is not destroyed or altered by sufficiently small perturbations. Usually, the stable waves are the ones that are observable. Because of the translational invariance, a stable wave, when perturbed, is allowed to settle at, possibly, another member of the family of its translations, and so the stability of waves is called orbital. If the wave is irrecoverably distorted by a perturbation, it is called unstable. Predicting whether the wave is unstable, and, if it is unstable, identifying the nature of the instability is important. The manifestations of instability vary. One way to distinguish different types of instability is to compare the growth of the perturbation to the rate at which it moves to an infinity. If, in the coordinate frame moving with the wave, perturbations to the wave are transported to infinity faster than they grow, the instability is called convective, as opposed to absolute instability when perturbations to the wave grow in time faster than they are moved away by the system. Absolute instability is characterized by the growth of perturbations at every point in the domain.

Traveling waves are ubiquitous in nature and human activities. They arise in applied problems from different fields. For example, in optical communication, traveling waves are bit carriers. In biomathematics, traveling waves appear in the form of calcium waves in tissue, to explain the nerve conduction, or as invasion fronts in population dynamics. In chemistry, the autocatalytic reactions exhibit periodic traveling waves and fronts. In botany, one may study traveling waves for the vegetation patterns or phyto-plankton blooms.

In combustion theory, traveling fronts are referred to as combustion fronts. The stability properties of combustion fronts may be given a multi-faceted interpretation. For example, it may be difficult to extinguish a stable combustion front because small perturbations of temperature and the amount of fuel may not be enough to disturb it. On the other hand, for the fire propagating along a fuse, one does not want small inhomogeneities of the surface or in the material of the fuse to affect the propagation of the fire, so considering stability properties of such fronts is a must.

In this book we concentrate on nonlinear traveling waves. Nonlinear traveling waves are solutions of nonlinear partial differential equations as opposed to linear equations. The theory of linear waves goes back to ancient times and is very well studied. Any introductory course on partial differential equations includes the study of the classical linear wave equation. The methods of finding the solutions to linear partial differential equations include Fourier analysis, separation of variables, and D'Alembert formula. The key feature of the linear equations is the superposition principle: a linear combination of the solutions of a linear partial differential equation is also a solution. Nonlinear equations do not enjoy the superposition property, so most of the techniques that work for linear equations are not useful in the nonlinear case. In this book we discuss examples of nonlinear equations that have traveling waves as solutions. Mostly, however, these equations do not have a closed form general solution; i.e. there is no formula for the general solution that one obtains by solving the equation. In the cases when the formula for the solution may not be obtained, one still may want to know whether non-trivial solutions such as traveling waves exist and whether these solutions are unique, stable, etc.

In the next sections we describe some of the classes of nonlinear partial differential equations that may support traveling wave solutions such as reaction-diffusion equations as well as some specific equations that are known to have traveling wave solutions.

1.2 REACTION-DIFFUSION EQUATIONS

The process of molecular diffusion is behind a variety of physical processes. For example, heat radiates from a heat source toward the cooler regions. Imagine a metal bar of length L such that L is much longer than the diameter of the bar. It may be associated then with a segment $[0, L]$ of a straight line, where 0 and L correspond to the ends of the bar and x $(0 < x < L)$ describes any other point of the bar. We assume that the temperature is the same for all points in the cross-section at the point x, and thus the temperature will be a function of only two independent variables: spatial variable x and time variable t. Let us denote the temperature of that bar at time t and location x by $u(x, t)$. If the initial temperature distribution and the temperature at the end-points are known, the evolution of the temperature $u(x, t)$ is given by the partial differential equation

$$u_t = d u_{xx}, \tag{1.1}$$

together with initial and boundary conditions on u. Here we use subscripts to denote a partial derivative, so u_t means the partial derivative of u with respect to t and u_{xx} is the second derivative of u with respect to x.

The equation (1.1) is called the heat equation. It is a very important equation in applied mathematics and is covered in every textbook on partial differential equations. The change in temperature here is assumed to be due to the diffusion of the heat only. Mathematically, the classical molecular diffusion is captured by the term proportional to the second derivative of u with respect to x. The coefficient of the proportionality $d > 0$, describes how fast or slow the diffusion occurs. It is called the diffusion coefficient or diffusivity. Generally speaking, the rate of diffusion does not have to be a constant, it may depend on the spatial variable x or time t, or even u itself, but in this text we assume that the diffusion rate d is the same everywhere and at all times.

In many natural phenomena, in addition to the diffusion, there are some other local processes going on. The combination of the diffusion and these processes is responsible for the evolution of the involved quantities. For example, let $u = u(x, t)$ be the concentration of some chemical at time $t > 0$ and at the location x. The chemical is produced or eliminated at the rate $f(u)$ which is called the reaction term or reaction kinetics. The rate of change in time of the concentration of this chemical can be written as

$$u_t = du_{xx} + f(u). \tag{1.2}$$

A partial differential equation such as (1.2) is called a reaction-diffusion equation. It is a partial differential equation of the second order. Recall that the order of a differential equation is the order of the highest derivative involved in it. Reaction-diffusion equations naturally arise in chemistry, therefore some of the chemical terminology is transferred to the mathematical theory of reaction-diffusion equations. The term $f(u)$ is called the reaction even if the underlying equation does not come from chemistry. Reaction-diffusion equations are ubiquitous in physics, biology, geology, and social sciences.

In a scalar equation like (1.2), the diffusion coefficient $d > 0$, without loss of generality, can be taken to be equal to 1. This is done by a "scaling transformation". By this we mean an affine transformation on the dependent and independent variables. In general, for the case of equation (1.2), we mean a transformation of the form

$$z = a_1 x + a_2, \quad \tau = b_1 t + b_2, \quad v = c_1 u + c_2, \tag{1.3}$$

where a_1, a_2, b_1, b_2, c_1, c_2 are constants. In the specific case where we want to use a transformation to take d to be equal to 1 in (1.2), we only need to rescale the spatial variable x. More precisely, let us make a change of variables $z = x/\sqrt{d}$, and not change the dependent variable u and the independent variable t. In other words, we take $a_1 = 1/\sqrt{d}$, $b_1 = c_1 = 1$ and $a_2 = b_2 = c_2 = 0$ in (1.3). By the chain rule then $u_x = u_z \frac{dz}{dx} = u_z/\sqrt{d}$, and furthermore, $u_{xx} = u_{zz}/d$. Replacing x with z in the equation (1.2), we obtain the equation

$$u_t = u_{zz} + f(u). \tag{1.4}$$

Another important phenomenon that is relatively easy to model is convection. As with the coefficient of the diffusion, the rate of the convection may be a function, but in this text, for simplicity, we assume that it is constant. Convection is modeled by a term proportional to the first spatial derivative of the unknown quantity u. Let the coefficient of proportionality be denoted by σ, then the term σu_x is added to the equation (1.2),

$$u_t = du_{xx} + \sigma u_x + f(u). \tag{1.5}$$

A partial differential equation such as (1.5) is called a reaction-convection-diffusion equation. Terms such as advection or drift or transport may be used instead of convection, depending on which aspect of the physical phenomenon is stressed, but mathematically they are described the same way. While the diffusion coefficient d needs to be nonnegative to make physical sense, the convection coefficient may be either positive or negative. Think of the equation (1.2) to be an equation that describes the evolution of a pollutant in a body of standing water such as a pond, while the equation (1.5) may be used to describe the evolution of the concentration of the pollutant in a river, with σu_x describing the transport of the pollution down the stream because of the water current.

Both equations (1.2) and (1.5) are examples of semilinear partial differential equations, which are equations that are linear in the highest order derivatives of the unknown function. These equations are simpler than the quasilinear partial differential equations where the highest derivatives may be multiplied by the lower derivatives or the unknown function itself or nonlinear equations which have terms nonlinear in the highest derivatives of the unknown function.

In the equations (1.2) or (1.5) the assumption that the time variable t is positive intuitively is clear. After all, the diffusion process is not reversible. A drop of paint in the water is not going to get back to its original shape after it dissolved because of the diffusion. The assumption that $t > 0$ is also a mathematical requirement. The equation (1.2) is a parabolic partial differential equation. The classical example of a parabolic partial differential equation is the heat equation (1.1). A "good" partial differential equation needs to be "well-posed". The solution should continuously depend on the parameters of the problem and on the initial conditions for the solution. Initial conditions which are in some sense close to each other should lead to solutions which are also close to each other and are not drastically different. The well-posedness guarantees nice properties of the solutions of the partial differential equation. The concept of the well-posedness is covered in any textbook on partial differential equations. For the heat equation and the reaction-diffusion equation (1.2) to be well-posed, we must require t to be positive.

In this text we will consider equations posed on a one-dimensional physical space by assuming that there is only one spatial variable x. This may seem as a technical simplification, but in reality, there are many applications when the geometry of the space is one dimensional. Think, for example, about a fire propagating along a fuse. The combustion reaction is confined to the fuse which, under reasonable assumptions, could be modeled as a segment of straight line. Moreover, in many cases the mathematical results obtained for a model posed on a one-dimensional physical space provide insight and adequately describe processes occurring in a two or higher-dimensional space. For example, there is a type of wave in equations posed on multidimensional space which are called planar waves. We show in Section 1.4 that for planar waves the existence of the wave is directly related to the existence of waves in a problem posed on one-dimensional physical space.

1.3 TRAVELING WAVES AS SOLUTIONS OF REACTION-DIFFUSION EQUATIONS

The stationary or time-independent equilibrium solutions of the equation (1.2) are found by setting the time-derivative equal to zero, so these solutions satisfy the equation

$$0 = du_{xx} + f(u). \tag{1.6}$$

This ordinary differential equation may have x-dependent solutions $u = u(x)$, which we call patterns. It may also have constant solutions that are independent of x, so called spatially homogeneous solutions or equilibria. If we are looking for spatially homogeneous or, simply, constant equilibria, the derivative with respect to the spatial variable x should be set to zero as well. Therefore, these are constant solutions of the algebraic equation

$$0 = f(u). \tag{1.7}$$

Assume that the equation (1.2) has at least one equilibrium. Then we may be interested if there are traveling wave solutions of this equation. Traveling wave solutions are functions of one variable $\xi = x - ct$ which represents the co-moving frame. So, say the traveling wave is given by $u = u(x,t) = \bar{u}(\xi(t,x)))$, where $\xi(t,x) = x - ct$, then by the chain rule we get

$$u_t = \frac{\partial \bar{u}(\xi(t,x))}{\partial t} = \frac{\partial \bar{u}(\xi(t,x))}{\partial \xi}\frac{\partial \xi}{\partial t} = -c\frac{\mathrm{d}\bar{u}(\xi(t,x))}{\mathrm{d}\xi}, \qquad (1.8)$$

since $\frac{\partial \xi}{\partial t} = -c$, and

$$u_x = \frac{\mathrm{d}\bar{u}(\xi(t,x))}{\mathrm{d}\xi}\frac{\partial \xi}{\partial x} = \frac{\mathrm{d}\bar{u}(\xi(t,x))}{\mathrm{d}\xi}, \qquad (1.9)$$

since $\frac{\partial \xi}{\partial x} = 1$. Furthermore,

$$u_{xx} = \frac{\partial}{\partial x}\left(\frac{\mathrm{d}\bar{u}(\xi(t,x))}{\mathrm{d}\xi}\right) = \frac{\partial}{\partial \xi}\left(\frac{\mathrm{d}\bar{u}(\xi(t,x))}{\mathrm{d}\xi}\right)\frac{\partial \xi}{\partial t} = \frac{\mathrm{d}^2\bar{u}(\xi(t,x))}{\mathrm{d}\xi^2}. \qquad (1.10)$$

Therefore the traveling wave solution satisfies the equation

$$0 = d\bar{u}'' + c\bar{u}' + f(\bar{u}), \qquad (1.11)$$

where the derivative is the ordinary derivative with respect to the variable ξ. It is a common practice to write the traveling wave equation (1.11) using the original variable u, understanding that a function of ξ does not look the same as the function of x and t and use $\bar{u} = \bar{u}(\xi)$ to denote a specific traveling wave the existence of which is known or assumed. In what follows we will pass from the partial differential equation (1.2) to the traveling wave equation

$$0 = du'' + cu' + f(u), \qquad (1.12)$$

without going through the calculations.

Another way to arrive to the same equation is as follows. To seek traveling waves we introduce in the equation (1.2) the moving coordinate $\xi = x - ct$, where the parameter c is not yet determined. We then perform the corresponding change of variables in the equation (1.2). If we replace x in $u = u(t,x)$ by ξ, we get a function $u(t,\xi + ct)$. Let us denote it by $\tilde{u}(t,\xi)$, then in (1.2), $f(u)$ will be directly replaced by $f(\tilde{u})$, but the partial derivatives must be treated using the chain rule for functions of two variables.

The chain rule says that for $g(\tau,\xi)$, $\tau = \tau(t,x)$, $\xi = \xi(t,x)$,

$$\frac{\partial g}{\partial t} = \frac{\partial g}{\partial \tau}\frac{\partial \tau}{\partial t} + \frac{\partial g}{\partial \xi}\frac{\partial \xi}{\partial t}, \qquad \frac{\partial g}{\partial x} = \frac{\partial g}{\partial \tau}\frac{\partial \tau}{\partial x} + \frac{\partial g}{\partial \xi}\frac{\partial \xi}{\partial x}. \qquad (1.13)$$

In our case the new variables are

$$\tau = \tau(t, x) = t, \qquad \xi = \xi(t, x) = x - ct, \qquad (1.14)$$

so

$$\frac{\partial \tau}{\partial t} = 1, \quad \frac{\partial \tau}{\partial x} = 0, \quad \frac{\partial \xi}{\partial x} = 1, \quad \frac{\partial \xi}{\partial t} = -c. \qquad (1.15)$$

We then obtain

$$\frac{\partial u(t, x)}{\partial x} = \frac{\partial \tilde{u}(t, \xi)}{\partial t} \frac{\partial t}{\partial x} + \frac{\partial \tilde{u}(t, \xi)}{\partial \xi} \frac{\partial \xi}{\partial x} = \frac{\partial \tilde{u}(t, \xi)}{\partial \xi},$$

$$\frac{\partial^2 u(t, x)}{\partial x^2} = \frac{\partial}{\partial x} \left(\frac{\partial \tilde{u}(t, \xi)}{\partial \xi} \right) = \frac{\partial}{\partial \xi} \left(\frac{\partial \tilde{u}(t, \xi)}{\partial \xi} \right) \frac{\partial \xi}{\partial x} = \frac{\partial^2 \tilde{u}(t, \xi)}{\partial \xi^2},$$

$$\frac{\partial u(t, x)}{\partial t} = \frac{\partial \tilde{u}(t, \xi)}{\partial t} + \frac{\partial \tilde{u}(t, \xi)}{\partial \xi} \frac{\partial \xi}{\partial t} = \frac{\partial \tilde{u}(t, \xi)}{\partial t} - c \frac{\partial \tilde{u}(t, \xi)}{\partial \xi}.$$

The equation (1.2) is then replaced by the equation

$$\tilde{u}_\tau - c \tilde{u}_\xi = d \tilde{u}_{\xi\xi} + f(\tilde{u}). \qquad (1.16)$$

It is a common practice to forgo replacing t with τ since they are identical and to drop the wave from $\tilde{u}$ to avoid introducing a new notation for the unknown function while understanding that the function u of the new variable ξ is related to the original function u of x and t but is not the same, thus obtaining the following version of (1.2) in the moving variable $\xi = x - ct$

$$u_\tau = d u_{\xi\xi} + c u_\xi + f(u). \qquad (1.17)$$

Now we recall that a traveling wave is a stationary solution when it is considered in the traveling coordinate. In other words, its profile does not depend on time $\tau = t$, and so we obtain

$$0 = d u'' + c u' + f(u). \qquad (1.18)$$

The equation (1.18) is the same equation as (1.12).

One may ask why we went through a deduction of (1.18), which is longer than the deduction of the equation (1.12). The answer is related to the further analysis of the traveling waves, more precisely, the analysis of their stability. The first step in the stability analysis is to consider the linear approximation of the underlying partial differential equation in a small neighborhood of the traveling wave. To do so, one rewrites the underlying partial differential equation, which is here the equation (1.2), in the coordinates (t, ξ), where $\xi = x - ct$ is a frame moving together with the wave. In this situation, we cannot assume that there is no dependence of the solutions on time other than through ξ, because the solutions near the wave generically do evolve with the time.

Another question that may arise is why we set $\xi = x - ct$ and not $x + ct$. There is no loss of generality in choosing one sign over the other in the notation of the coordinate frame. We could go through the same procedure with $\xi = x + ct$ and as a result we would get an equation like (1.12) but with a negative sign in front of the term cu'. Since we never specified what the sign of c, it is not restrictive at all.

The sign of the parameter c is another topic of discussion. First of all, let us mention that c is allowed to be 0. In this case the corresponding solution is called a standing wave. Standing waves fall in the category of solutions that we call patterns. Assume now that we know or were able to prove that a traveling front $u = u(\xi)$ exists as a solution of (1.2) for some $c = c_0 > 0$. As a front u has two different rest states A and B, which are constant equilibria of the equation (1.2) and as such are found as zeros of $f(u)$: $f(A) = f(B) = 0$. In other words, u solves (1.18) with $c = c_0$,

$$0 = du'' + c_0 u' + f(u), \tag{1.19}$$

and satisfies boundary-like condition

$$\lim_{\xi \to -\infty} u(\xi) = A, \quad \lim_{\xi \to +\infty} u(\xi) = B. \tag{1.20}$$

We call this conditions "boundary-like" instead of "boundary" because they are imposed as limits instead of values of the function u at some fixed values of ξ.

Now let us check if the equation (1.2) has a front in the case when $c = -c_0 < 0$, in other words, if the equation (1.18) has a solution $u = u(\xi)$ when $c = -c_0$,

$$0 = d\frac{d^2 u}{d\xi^2} - c_0 \frac{du}{d\xi} + f(u). \tag{1.21}$$

We introduce a new variable $z = -\xi$ in the equation (1.21) and notice that $\frac{du}{dz} = -\frac{du}{d\xi}$ and, thus, $\frac{d^2 u}{dz^2} = \frac{d^2 u}{d\xi^2}$. The equation (1.21) then in terms of z reads

$$0 = d\frac{d^2 u}{dz^2} + c_0 \frac{du}{dz} + f(u). \tag{1.22}$$

The conditions (1.20) read

$$\lim_{z \to -\infty} u(\xi) = B, \quad \lim_{z \to +\infty} u(\xi) = A. \tag{1.23}$$

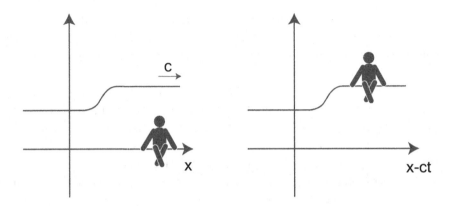

Figure 1.2 This figure illustrates passing to the moving with the front coordinate frame, in which the front becomes stationary.

The sign of c in ξ plays the role of the direction of the movement. Assume that the front moves to the right, or from $-\infty$ to ∞ along the x-axis. Visualize the moving front and fixed axis, like a moving train and the platform. You are viewing this system while standing on the platform. Now switch your reference frame: imagine that you are riding on the train. The platform will be moving to the left. So if you are in co-moving with the front frame, the points on the axis move towards $-\infty$ as time progresses, thus, $\xi = x - ct$, $c > 0$. The situation is illustrated in Figure 1.2.

When we consider a negative velocity, $-c_0$, the direction of the movement is switched to the one from the right to the left, but the rest states A and B are switched too, so it is exactly the same solution. The situation may be visualized as reflecting the horizontal ξ-axis and thus the solution with respect to the vertical u-axis. We conclude that it is sufficient to consider traveling wave equations for $c > 0$.

The situation is similar with pulses: it is enough to consider c either positive or negative. No new solutions appear if the sign of c is changed to the opposite.

1.4 PLANAR WAVES

Let us now discuss reaction-diffusion equations posed on two-dimensional physical space. In other words, we consider a reaction-diffusion equation with the following independent variables: the time $t > 0$, and spatial variables x_1, $x_2 \in \mathbb{R}$. The reaction-diffusion equation reads

$$u_t = d(u_{x_1 x_1} + u_{x_2 x_2}) + f(u). \tag{1.24}$$

In (1.24) it is assumed here that the diffusion is uniform in both directions. Typically, in the literature, an alternative notation may be used in this equation which uses the Laplacian operator Δ : $\Delta u = u_{x_1 x_1} + u_{x_2 x_2}$.

We want to identify conditions under which there exists a front solution that propagates in a preferred direction, say along x_2-coordinate. We pass in the equation (1.24) to the new coordinate $\xi_2 = x_2 - ct$ and keep t and x_1 without a change. The equation (1.24) becomes

$$u_t = d(u_{x_1 x_1} + u_{\xi_2 \xi_2}) + c u_{\xi_2} + f(u). \tag{1.25}$$

You may notice that we denoted by u a function of different variables than originally, in (1.24).

Assume that there exists a traveling wave solution $\tilde{u}(\xi_2)$ of the associated equation

$$u_t = d u_{\xi_2 \xi_2} + c u_{\xi_2} + f(u), \tag{1.26}$$

which we obtained by dropping the derivative with respect to x_2. The same function solves the equation (1.25). The front $u(t, x_1, x_2) = \tilde{u}(x_2 - ct)$ is called a planar wave solution of the equation (1.25). Geometrically, it may be visualized as a surface formed by a continuum of one-dimensional profiles $\tilde{u}(x_2 - ct)$ stack up perpendicularly to x_1 that are moving all together in the direction of x_2 with the velocity c. See Figure 1.3 for illustration.

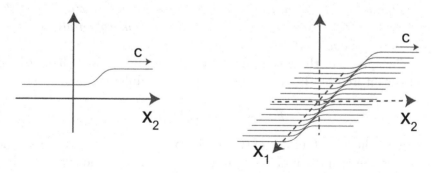

Figure 1.3 This figure illustrates the relation between a one-dimensional wave and a planar wave that is associated with it.

The analysis of the existence of the planar traveling waves in (1.25) is reduced to the analysis of the existence of traveling waves in the associated one-dimensional version (1.26). In some situations, the stability analysis of a planar front is also reduced to the stability analysis of the associated one-dimensional front [37, 59].

We finish this section with few examples of planar fronts in applications. In a dry grass or forest fire, combustion fronts may be propagating as planar fronts. A solidification front that develops in coating materials is used to improve the durability of the repairs of damaged components of deteriorated steam circuit parts of thermal power stations [18]. When foraging, locusts travel in a planar front toward the available crop as they leave behind destroyed crops [10].

1.5 EXAMPLES OF REACTION-DIFFUSION EQUATIONS

1.5.1 Fisher-KPP equation

Fisher-KPP equation appears in models from many different disciplines such as biology and ecology and appears to be a prototype equation that captures the so-called spreading phenomena.

In 1937, R.A. Fisher[1] published a paper "The wave of advance of advantageous gene". The goal of this paper was to "examine the properties of the differential equation determining the gene spread in the simplest case." The suggested equation was formulated for a one-dimensional spatial variable. Here is a quote from the paper that describes the setting of the problem:

Consider a population distributed in a linear habitat, such as a shore line, which it occupies with uniform density. If at any point of the habitat a mutation occurs, which happens to be at some degree, however slight, advantageous to survival, in the totality of its effects, we may expect the mutant gene to increase at the expense of the allelomorph or allelomorphs previously occupying the same locus.

The spread of the advantageous gene occurs in a wave-like fashion and is captured by the scalar nonlinear partial differential equation

$$p_t = dp_{xx} + mp(1 - p), \tag{1.27}$$

where p, which is a function of the spatial variable x and the time variable $t > 0$, is the frequency of the mutant gene, $m > 0$ is the constant that

---

[1]R.A. Fisher (1890–1962) was a British statistician, mathematician, geneticist, and biologist.

represents the intensity of the selection in favor of the mutant gene, and $k > 0$ is the "coefficient of diffusion analogous to that in physics". The quantity $1 - p$ in the reaction term $p(1 - p)$ represents the parent allelomorph of the mutant gene. One can interpret the term $p(1 - p)$ as a chance encounter between the carrier of the mutant gene with a non-carrier of the mutant gene.

This equation has two spatially homogeneous equilibrium points $p = 0$ and $p = 1$. The equilibrium $p = 0$ corresponds to the state when no mutant gene is present, and the equilibrium $p = 1$ describes the state when the mutant gene took over the population. Let us recall that the term equilibrium means that the solution is time (t) independent, and "spatially homogeneous" means that the solution is independent of the spatial variable x, therefore the term "a spatially homogeneous equilibrium point" simply means a constant solution. To find such solutions, we set in the equation (1.27) the derivatives of the solution with respect to the time t and the space x to be zero and solve the algebraic equation $0 = mp(1 - p)$ for p.

The same results were obtained independently by a group of Russian researchers A. Kolmogorov[2], I. Petrovsky[3], and N. Piskunov[4]. Their paper "Investigation of the equation of diffusion combined with increasing of the substance and its application to a biology problem" was also published in 1937.

An alternative form of the equation (1.27) that is often used in the literature is

$$u_t = du_{xx} + mu \left(1 - \frac{u}{K} \right). \tag{1.28}$$

Here $d > 0$ and $K > 0$ are constant parameters. When u is a population of any type, K is interpreted as the carrying capacity of the environment. It is easy to see that the equations (1.27) and (1.28) are equivalent. Equations are said to be equivalent if given a solution of one of these equations, the solution of the other one is known too. The relation between the solutions of (1.31) and (1.27) is simply $u(x, t) = Kp(x, t)$. To

---

[2]Andrey Kolmogorov (1903–1987) was a Russian mathematician known for his contributions to probability theory, topology, turbulence studies, mathematical analysis, Kolmogorov complexity, and the KPP equation, among other topics.

[3]Ivan Petrovsky (1901–1973) was a Russian mathematician who was focused on partial differential equation analysis.

[4]Nikolai Piskunov (1908–1977) was a Russian mathematician and educator working in the area of partial differential equations.

see that, replace $u(x,t)$ in the equation (1.31) with $Kp(x,t)$ and divide both sides of the equation (1.31) by K.

Sometimes another equivalent form of this equation is also used

$$u_t = du_{xx} + Ru(K - u). \tag{1.29}$$

Equation (1.29) is the same as (1.31) with the parameter m replaced by the new parameter $R = m/K$. The equation (1.29) is related to another famous equation. Indeed, if we assume that the population u is uniformly distributed in x and one is only concerned by the time evolution of the population at a fixed location, the second derivative of u with respect to the spatial variable x must be ignored. The equation therefore is reduced to the logistic equation

$$u_t = Ru(K - u), \tag{1.30}$$

which is an ordinary differential equation with one independent variable t. The logistic equation is one of the cornerstone equations in population modeling that captures the concept of population dynamics of which is influenced by the carrying capacity of the environment. It was first studied by P.F. Verhulst[5] in the 1800's as the equation which predicts human population growth in different countries.

The equation (1.27) has two constant parameters, and the equations (1.28) and (1.29) have three constant parameters. As it is shown above, these equations are equivalent. The equation

$$u_t = u_{xx} + u(1 - u) \tag{1.31}$$

does not have any parameters. It is also equivalent to the equations (1.27), (1.28), and (1.29), as long as the parameters in these equations are strictly positive. Let us show that the equations (1.31) and (1.27) are equivalent. We divide the equation (1.27) by m and rescale the time and space variables as

$$z = \sqrt{\frac{m}{d}}x, \quad \tau = mt. \tag{1.32}$$

In the new variables the equation (1.27) becomes (1.31). We note that the equations (1.31) and (1.27) are related through the rescaling of the

---

[5]Pierre François Verhulst (1804–1849) was a Belgian mathematician and biologist known for his contributions to population modeling.

independent variables, while the relation to the equations (1.28) and
(1.29) also includes rescaling the unknown function.

The equation (1.31) is a famous example of a scalar reaction-diffusion
equation that describes the propagation of waves, more precisely, the
propagation of fronts. The proof of the existence of some front solutions
in this equation is given in Section 3.3.

There is a more general equation that shares all of the important
properties of the Fisher-KPP equation (1.31) which is called the generalized Fisher-KPP equation. The generalized Fisher-KPP equation is
given by

$$u_t = u_{xx} + f(u), \tag{1.33}$$

where the reaction term $f(u)$ satisfies the following conditions [26, 65].
There are exactly two values, say $u = 0$ and $u = K$, such that $f(0) = 0$
and $f(K) = 0$, which means that the equation has two constant equilibrium points $u = 0$ and $u = K$. In addition, there are assumptions on
the derivatives of f,

$$f'(0) > 0, \quad f'(K) < 0, \quad f''(u) < 0. \tag{1.34}$$

These conditions of f and its derivative guarantee the existence of front
solutions.

Assuming that one of the equilibrium solutions is $u = 0$ is not restrictive. Let us say, that $f(u) = 0$ has two different solutions $u = K_1$
and $u = K_2$. We then make a substitution $v = u - K_1$ in the equation
(1.31) and notice that $u_t = v_t$ and $u_{xx} = v_{xx}$, since K_1 is a constant.
The function $f(u)$ becomes $\tilde{f}(v) = f(v + K_1)$ and so

$$\tilde{f}(0) = 0, \quad \tilde{f}(K_2 - K_1) = f(K_2 + (K_1 - K_2)) = f(K_2) = 0. \tag{1.35}$$

The equation

$$v_t = k v_{xx} + \tilde{f}(v), \tag{1.36}$$

is a Fisher-KPP equation with constant solutions $u = 0$ and $u = K_2 - K_1$,
where we can denote $K_2 - K_1 = K$. It is a usual practice to assume in a
reaction-diffusion equation where the reaction term is not specified that
one of the constant solutions is 0.

It is known that the equation (1.31) has fronts that asymptotically
connect the equilibrium $u = 1$ to the equilibrium $u = 0$. More precisely,
with respect to the moving coordinate $\xi = x - ct$ with $c > 0$, there exist
solutions $u = u(\xi)$ of the associated traveling wave equation

$$0 = \frac{\mathrm{d}^2 u}{\mathrm{d}\xi^2} + c\frac{\mathrm{d}u}{\mathrm{d}\xi} + u(1 - u) \tag{1.37}$$

with boundary-like conditions

$$\lim_{\xi \to -\infty} u(\xi) = 1, \quad \lim_{\xi \to +\infty} u(\xi) = 0, \tag{1.38}$$

for any $c > 0$. Recall that this means that for each $c > 0$, there is a family of fronts which are translations of each other. Not all of these waves are physical and therefore not all of them are relevant. For example, in biological applications, only nonnegative values of u have a meaning, so if a solution turns negative at any point, it is not realistic. In the case of the equation (1.31), there is a value c^* of the velocity c such that the fronts that move with speeds less than c^* do not have a realistic physical interpretation, but the fronts that move with speed $c \geq c^*$ are meaningful.

Some closed form solutions of the equation (1.37) are known. For example, when $c = 5/\sqrt{6}$, a function

$$u = 1 - \left(1 + \frac{a}{\sqrt{6}} e^{\xi/\sqrt{6}}\right)^{-2}, \tag{1.39}$$

solves the equation for an arbitrary constant a. This solution was found in [1] using a quite complicated analytical technique. In Section 3.3 we show how to prove the existence of the traveling fronts without actually solving the equation, but, instead, by using a slope field analysis.

1.5.2 Nagumo equation

Nagumo equation was originally formulated in the context of the modeling of the propagation of an electrical impulse along an axon. The process was originally explained by English physiologists and biophysicists Alan Lloyd Hodgkin and Andrew Fielding Huxley in 1952 in their celebrated work on the modeling action potential propagation in the squid's giant axon. The model that captures key features of this process is called Hodgkin-Huxley model.

Hodgkin and Huxley later, in 1963, received the Nobel Prize in Physiology or Medicine "for their discoveries concerning the ionic mechanisms involved in excitation and inhibition in the peripheral and central portions of the nerve cell membrane" [45]. We refer the reader to the Award Ceremony speech [39], which explains with a good level of detail both the experiments and the explanations.

The Hodgkin-Huxley model consists of several coupled ordinary differential equations. The unknown functions in the system are functions of time only. They are associated with the activation of the potassium and sodium channels, the inactivation of the sodium channel, and the voltage. The equations are nonlinear and the system is difficult to analyze.

Some simplifications to the system were introduced. Related models were suggested by the American biologist Richard FitzHugh [27] in 1961. Another, simpler model was formulated by Japanese physicists J. Nagumo, S. Arimoto, and S. Yoshizawa in [80] in 1962 where the authors used rules of an electric circuit to model the electricity conduction in an axon. The model became known as the FitzHugh-Nagumo model. The model consists of coupled partial differential equations that describe the evolution of the membrane potential or the voltage along the axon of a single neuron and the recovery variable that captures the mechanism of resetting the membrane. This system and some of the solutions are discussed in Section 3.3.1.

In these models the solution of interest had a form of a pulse, which is a solution that asymptotically "originates" and "ends" in the same constant state while propagating in some direction. In a later paper [81] in 1965, Nagumo and collaborators created an equation that captures signal propagation, which is a form of a front. A front asymptotically "originates" at one constant state and "ends" in a different same constant state in some moving frame. These constant states that are the limits of the solution at $\pm\infty$ are sometimes called the rest states of the front or the pulse.

The Nagumo equation is the prototype equation that was found to support traveling fronts. The equation reads

$$Cv_\tau = \frac{1}{r}v_{zz} + a(v_1 - v)(v - v_2)(v - v_3), \tag{1.40}$$

where a, C and r and $v_1 < v_2 < v_3$ are positive constants. It is easy to see that $v = v_1$, $v = v_2$ and $v = v_3$ correspond to constant solutions or spatially homogeneous equilibria of the equation (1.40). Equation (1.40) can be rescaled and cast in the form

$$u_t = u_{xx} + u(1 - u)(u - a), \tag{1.41}$$

where $a \in (0, 1)$. For example, one can take a rescaling

$$u = \frac{v - v_1}{v_3 - v_1},$$

which transforms $v = v_1$ into $u = 0$ and $v = v_3$ into $v = 1$, then make a change of the independent variables $\tau = \frac{a}{C}t$ and $z = \sqrt{ar}x$.

It is easy to see that the equation (1.41) has three constant solutions: $u = 0$, $u = 1$, and $u = a$. Out of these solutions, $u = 0$ and $u = 1$ are stable as solutions of the equation (1.41), which means that small perturbations to these solutions decay in time. We shall discuss the stability of the solutions later, in Chapter 4, but here we want to mention that the equation that has two stable constant states is called bistable.

Of particular interest are front solutions of the equation (1.41) that have these stable constant solutions as limits. It is known that this equation has traveling fronts that depend on the value of a. More precisely, a front $u = u_0(\xi)$, where $\xi = x - ct$ represents the moving coordinate, satisfies the following boundary-like conditions:

$$
\begin{aligned}
\lim_{\xi \to -\infty} u_0(\xi) = 1, \ \lim_{\xi \to +\infty} u_0(\xi) = 0, \ \text{when } 0 < a < 1/2, \\
\lim_{\xi \to -\infty} u_0(\xi) = 0, \ \lim_{\xi \to +\infty} u_0(\xi) = 1, \ \text{when } a > 1/2.
\end{aligned}
\tag{1.42}
$$

When $a = 1/2$, the system has standing fronts connecting the equilibria $u = 0$ and $u = 1$.

Bistable equations such as the Nagumo equation (1.41) play an important role in understanding of mechanisms of pattern formation. They have been studied extensively. Fundamental results that help to understand pattern forming mechanisms in reaction-diffusion equations were obtained in [24, 25]. These results play an important role in proving the existence and understanding stability properties of more complex solutions of Fitzhugh–Nagumo equation (see, for example, [16, 43, 57, 68, 82] and references therein).

Moreover, there is a more general equation that shares all properties of the Nagumo equation (1.41). The equation is called the generalized Nagumo equation. It is an equation of the following form

$$
u_t = u_{xx} + f(u),
\tag{1.43}
$$

provided the nonlinearity $f(u)$ is a smooth function that has zeros at $u = 0$, $u = 1$, and $u = a$, where $a \in (0, 1)$, i.e., $f(0) = f(1) = f(a) = 0$, and $f'(0) < 0$, $f'(1) < 0$. Here again we assume that the constant solutions occur at 0 and 1 without loss of generality, since the variable u may be appropriately scaled to have this property. Stability of the constant states $u = 0$ and $u = 1$ as solution of the equation (1.43) is encoded in the conditions on the derivative of the nonlinearity $f(u)$ at these points.

1.6 OTHER PARTIAL DIFFERENTIAL EQUATIONS THAT SUPPORT WAVES

1.6.1 Nonlinear diffusion, convection, and higher order derivatives

Reaction-diffusion equations are not the only partial differential equations that support traveling waves. Many equations that support traveling waves are of order higher than two and many equations are not semilinear. We here mention two important equations: the Burgers equation and the Korteweg-de Vries (KdV) equation. Both of these equations have nonlinear convection terms which means that the rate of the convection is not a constant, but is a function that depends on the unknown quantity that the equations are describing.

Burgers equation has applications in mechanics and gas dynamics. It is a second-order equation with a nonlinear convection term. The KdV equation is known as the "shallow water equation". It is a partial differential equation of the third order with a nonlinear convection term. These equations are not semilinear because there are nonlinear terms that involve derivatives, but both equations are classified as quasilinear, because the highest derivatives participate in these equations in a linear manner.

We discuss these equations in more detail below.

1.6.2 Burgers equation

Burgers equation was originally formulated in relation to fluid mechanics by H. Bateman[6] [5] in 1915, and studied in 1948 by J.M. Burgers[7] as a "mathematical model illustrating the theory of turbulence" [14]. Later the equation was found to be applicable in gas dynamics, acoustics, and other areas of science. The equation is

$$u_t = du_{xx} - uu_x. \qquad (1.44)$$

It describes the velocity $u(t,x)$ of a viscous fluid. Here $t > 0$ is the time variable, x is the spatial variable, and d is the diffusion parameter. One-dimensional spatial configuration is used, for example, to model a fluid flowing in a thin long pipe.

---

[6]Harry Bateman (1882–1946) was an English mathematician mostly known for his work on partial differential equations of mathematical physics and special functions.

[7]Johannes Martinus Burgers (1895–1981) was a Dutch physicist and mathematician who is the namesake of the Burgers' equation, the Burgers vector in dislocation theory, and the Burgers material in viscoelasticity.

The equation (1.44) is called a viscous Burgers equation if $d > 0$. It is called as a quasilinear partial differential equation of the second order with a nonlinear convection term. It has traveling waves as solutions. There is also an inviscid Burgers equation which is equation (1.22) with $d = 0$, and so it is a nonlinear equation of the first order,

$$u_t = -uu_x. \tag{1.45}$$

The latter is an example of a conservation law. From the mathematical point of view, a scalar conservation law is an equation of the form

$$u_t + \frac{\partial f(u)}{\partial x} = 0. \tag{1.46}$$

It is easy to see that the inviscid Burgers equation is an equation of the form (1.46) with $f(u) = u^2/2$.

The inviscid Burgers equation with given initial conditions may be solved using the method of characteristics. It may have solutions that develop a discontinuity in time. These types of solutions are called shock waves. There is an area of mathematical research focused on conservation laws and shock waves.

1.6.3 Korteweg-de Vries (KdV) equation

It is important for engineering progress to understand the water waves. The famous reference to stable, i.e. shape-preserving, waves was made by J. S. Russell[8] in 1845. He was conducting experiments aimed at designing a boat that could be effectively used in canals. In his experiments, he observed a wave that he called the "Wave of Translation," which is now called Russell's solitary wave.

Korteweg-de Vries equation or, for short, KdV equation, is an equation that models waves in shallow water that are generated by the gravitational force. Let $v = v(\tau, z)$ be the elevation of the water relative to the horizontal bottom at a spatial location z at the time τ, then v satisfies the following equation

$$v_\tau + \frac{h^2}{6}\sqrt{gh}v_{zzz} + \sqrt{gh}v_z + \frac{3}{2}\sqrt{\frac{g}{h}}vv_z = 0, \tag{1.47}$$

where g is the gravitational constant and h is the water depth. This equation first appeared as a footnote in the Essay On The Theory of Flowing

---

[8]John Scott Russell (1808–1882) was a Scottish naval engineer and architect best known for his work on ship design and fluid dynamics.

Water by J. Bousinesq[9]. The equation is named after D. Korteweg[10] and his doctoral student G. de Vries[11] who rediscovered the equation in 1895 [66].

To obtain a version of this equation that does not depend on the units of measurement, in other words, a dimensionless equation, one introduces new variables in a way that "cancels" the units. Here we divide v which is measured in meters by h, which is also measured in meters $u = \frac{v}{h}$ and rescale the time by setting $t = \frac{1}{6}\sqrt{\frac{g}{h}}\tau$ (check the units of the expression on the right of this formula as an exercise). To eliminate the stand alone first-order derivative in z, we further replace the spatial variable z with a new spatial variable $x = \frac{z}{h} - t$ and obtain

$$u_t + u_{xxx} + 6uu_x = 0. \tag{1.48}$$

The KdV equation is an example of a class of equations called integrable. Essentially, an integrable equation is an equation for which large classes of solutions can be found (see [19] for a book on the topic of integrable equations). For the KdV equation a traveling wave may be obtained in a closed form, which means there is a formula that describes this wave. Before providing this formula, we introduce a moving coordinate $\xi = x - ct$ in the equation (1.48) and obtain

$$u_t - cu_\xi + u_{\xi\xi\xi} + 6uu_\xi = 0. \tag{1.49}$$

The corresponding traveling wave equation which is an ordinary differential equation is

$$-c\frac{du}{d\xi} + \frac{d^3u}{d\xi^3} + 6u\frac{du}{d\xi} = 0. \tag{1.50}$$

It is not difficult to check by substitution that, for every $c > 0$,

$$u_0 = \frac{c}{2}\text{sech}^2\left(\frac{\sqrt{c}}{2}(\xi - \xi_0)\right) \tag{1.51}$$

is a solution of the equation (1.50) for any ξ_0.

---

[9]Joseph Boussinesq (1842–1929) is a French mathematician and physicist known for his contributions to the theory of hydrodynamics, vibration, light, and heat. In particular, he developed mathematical theory to explain experimental observations of the solitons made by Russel.

[10]Diederik Korteweg (1848–1941) was a Dutch physicist and mathematician best known for his work on the KdV equation.

[11]Gustav de Vries (1866–1934) was a Dutch mathematician who worked on the KdV equation together with Korteweg.

The expression (1.51) is also quite straightforward to deduct. First we notice that $2u\frac{du}{d\xi} = \frac{d(u^2)}{d\xi}$, then the equation (1.50) may be integrated once

$$-cu + \frac{d^2 u}{d\xi^2} + 3u^2 = C, \tag{1.52}$$

where C is the constant of integration. From the equation (1.49) it is obvious that any constant solution is an equilibrium. Assume that we are interested in positive solutions that approach 0 at $+\infty$, the constant of integration C in (1.52) then must be 0,

$$-cu + \frac{d^2 u}{d\xi^2} + 3u^2 = 0. \tag{1.53}$$

We multiply the equation (1.53) by u_ξ,

$$-cu\frac{du}{d\xi} + \frac{d^2 u}{d\xi^2}\frac{du}{d\xi} + 3u^2\frac{du}{d\xi} = 0 \tag{1.54}$$

and integrate to obtain

$$-\frac{c}{2}u^2 + \frac{1}{2}\left(\frac{du}{d\xi}\right)^2 + u^3 = C, \tag{1.55}$$

where C is again a constant of integration. The same assumption on the limit at $\xi \to +\infty$ leads to the equation

$$-\frac{c}{2}u^2 + \frac{1}{2}\left(\frac{du}{d\xi}\right)^2 + u^3 = 0. \tag{1.56}$$

Solving this equation for the derivative leads to

$$\frac{du}{d\xi} = \pm\sqrt{cu^2 - 2u^3}. \tag{1.57}$$

The equation (1.57) is a separable ordinary differential equation and it can be written as

$$\frac{du}{u\sqrt{c - 2u}} = \pm d\xi, \tag{1.58}$$

where we used the assumption that the solution is positive. The left hand side of (1.58) may be integrated using substitution $u = \frac{c}{2}\operatorname{sech}^2(\tilde{u})$, since then

$$c - 2\tilde{u} = c\tanh^2(\tilde{u}) \quad \text{and} \quad d\tilde{u} = -c\operatorname{sech}^2(\tilde{u})\tanh(\tilde{u})d\tilde{u}, \tag{1.59}$$

and so, after some simplifications, we get

$$-\int \frac{2d\tilde{u}}{\sqrt{c}} = \pm \int d\xi, \qquad (1.60)$$

or

$$\tilde{u} = \pm \frac{\sqrt{c}}{2}\xi + C. \qquad (1.61)$$

The latter implies that

$$u = \frac{c}{2}\operatorname{sech}^2\left(\frac{\sqrt{c}}{2}\xi + C\right), \qquad (1.62)$$

where we used that sech is an even function and C is an arbitrary constant without a prescribed sign. We denote $\xi_0 = \frac{2C}{c}$ and get the solution (1.51) in the form reflecting the translational invariance of the equation. Since the limits of this solution as $\xi \to \pm\infty$ are both 0, so the solution has a form of a pulse. Thus, for any $c > 0$ and ξ_0

$$u_0 = \frac{c}{2}\operatorname{sech}^2\left(\frac{\sqrt{c}}{2}(\xi - \xi_0)\right) \qquad (1.63)$$

is a traveling pulse of the equation (1.49). Notice that not only 0 but any constant is an equilibrium for (1.49). But since u represents the elevation of the water, it makes no sense to consider negative constant solutions.

Further generalizations of (1.63) also exist, based on a nonlinear superposition principle, discussion of which we skip, but which basically means that if we know one solution to the equation, we can build families of solutions of the equation.

The formula (1.63) does not describe all of the traveling waves in (1.49). For example, the restriction $c > 0$ implies that this formula captures waves that move to the right in the chosen coordinate frame. The formula for the waves that move to the left $(c < 0)$ is also known.

The solutions of the KdV equation of the form (1.63) are called solitons. Solitons were discovered by N. Zabusky[12] and M. Kruskal[13]

---

[12]Norman J. Zabusky (1929–2018) was an American computational physicist who is known for his discovery of the soliton in the KdV equation (with Kruskal), and for his pioneering idea to use computer simulations to gain insights into nonlinear differential equations and the importance of visualization as a tool in analyzing complex fluid dynamical and wave systems

[13]Martin David Kruskal (1925–2006) was an American applied mathematician and physicist known for his discovery of solitons (with Zabusky) and contributions to plasma physics, inverse scattering transform, among other topics. He also developed the Kruskal Coordinates, used in the theory of relativity to explain black holes.

in 1965 in the paper [105] where they numerically integrated the KdV equation and observed that solutions decompose in a sequence of single-peak waves which did not change as a result of a collision with other solitons. This property is what makes a soliton special. A soliton may be defined as a traveling pulse that preserves its shape and velocity after interacting with another soliton, with the exception of a possible phase shift.

As in the case of Fisher-KPP and Nagumo equations, a generalized version of the KdV equation (1.48) also exists. The generalized KdV equation [87] is

$$u_t + u_{xxx} + u^p u_x = 0, \quad p > 0. \tag{1.64}$$

This equation shares general properties of the KdV equation (1.48) and has the following solution

$$u_0 = \left[\frac{c(p+1)(p+2)}{2} \right]^{1/p} \operatorname{sech}^{2/p} \left(\frac{p\sqrt{c}}{2} (\xi - \xi_0) \right). \tag{1.65}$$

CHAPTER 2

Systems of Reaction-Diffusion Equations posed on infinite domains

2.1 SYSTEMS OF REACTION-DIFFUSION EQUATIONS

In Chapter 1 we discussed the equation

$$u_t = du_{xx} + f(u) \tag{2.1}$$

that describes the evolution of one quantity u. In many situations the phenomena that we are trying to describe mathematically are driven by interactions of more than one quantity. For example, in combustion theory many processes are driven by the concentration of the fuel and the temperature, both as functions of time and space. Sometimes, like in porous media combustion, one has to consider the influence of the pressure too. So the models for these processes must include more than one equation. In mathematical biology, predator and prey relations also are often modeled using more than one equation, coupled to each other. In a form consistent with (2.1), we can write a system as

$$\mathbf{u}_t = D\mathbf{u}_{xx} + \mathbf{f}(\mathbf{u}), \tag{2.2}$$

DOI: 10.1201/9781003147619-2

where the bold face letters denote vectors: $\mathbf{u} = (u_1(t,x), ..., u_n(t,x))^T$, $\mathbf{f} = (f_1, ..., f_n)$, where each f_i is a scalar function of u_1, ..., u_n. The quantities u_i do not necessarily diffuse at the same rate, so we assume that $D = \text{diag}(d_1, d_2, \ldots, d_n)$ is a diagonal matrix of size $n \times n$ with positive entries. More precisely, the equation (2.2) is the matrix form of the system

$$
\begin{aligned}
\frac{\partial u_1}{\partial t} &= d_1 \frac{\partial^2 u_1}{\partial x^2} + f_1(u_1, ..., u_n), \\
\frac{\partial u_2}{\partial t} &= d_2 \frac{\partial^2 u_2}{\partial x^2} + f_2(u_1, ..., u_n), \\
&\cdots \\
\frac{\partial u_n}{\partial t} &= d_2 \frac{\partial^2 u_n}{\partial x^2} + f_n(u_1, ..., u_n).
\end{aligned}
\tag{2.3}
$$

The spatially homogeneous steady states of this system are the solutions of the algebraic system of equations

$$
\begin{aligned}
f_1(u_1, ..., u_n) &= 0, \\
f_2(u_1, ..., u_n) &= 0, \\
&\cdots \\
f_n(u_1, ..., u_n) &= 0,
\end{aligned}
$$

if any exist. The traveling waves are vector-functions $\mathbf{u}$ or, equivalently, sets of functions u_1,..., u_n that in the co-moving frame $\xi = x - ct$ satisfy the following system of ordinary differential equations

$$
\begin{aligned}
d_1 \frac{d^2 u_1}{d\xi^2} + c \frac{du_1}{d\xi} + f_1(u_1, ..., u_n) &= 0, \\
d_2 \frac{d^2 u_2}{\partial \xi^2} + c \frac{du_2}{d\xi} + f_2(u_1, ..., u_n) &= 0, \\
&\cdots \\
d_n \frac{d^2 u_n}{d\xi^2} + c \frac{du_n}{d\xi} + f_n(u_1, ..., u_n) &= 0,
\end{aligned}
\tag{2.4}
$$

which in matrix notations reads

$$
D \frac{d^2 \mathbf{u}}{d\xi^2} + c \frac{d\mathbf{u}}{d\xi} + \mathbf{f}(\mathbf{u}) = \mathbf{0},
\tag{2.5}
$$

where $\mathbf{0}$ is the zero vector with n components.

Notice that the system (2.3) is a system of parabolic partial differential equations as long as all of d_i's are positive. Because of the parabolicity, these equations have some nice properties: they are well posed. In many applications, some of the quantities diffuse and others do not diffuse. For example, in the systems describing combustion of solid fuels, the fuel does not diffuse, but the temperature in the burning zone does. In biomath, one may want to consider a "predator-prey model" where the prey does not move or moves much slower than the predator. In the latter case, it makes sense to set the diffusion coefficient in front of the prey variable to be zero. For example, this would be the case in a model for locust feeding on the crops. In modeling for a malignant tumor growth, malignant cells are mobile and healthy cells do not spread [74].

Mathematical models in these cases may still be written in a form (2.2) or (2.3), but in the matrix D some of the entries may be zero. When some but not all of the diffusion coefficients are zero, the system is called partly parabolic. (In the literature these type of systems are also called partly dissipative [92, p.283] or partially degenerate [64]).

For proving the existence of traveling waves in partly parabolic systems, compared to parabolic reaction-diffusion systems, no different techniques are required. Moreover, if the existence is proved using applied dynamical system approach, then sometimes replacing a relatively small diffusion coefficient with zero is justified and convenient because the resulting dynamical system is lower dimensional. Indeed, the first step in this method is to rewrite (2.4) as a system of ordinary differential equations of the first order. Each equation in (2.4) is then replaced with two equations. For example,

$$d_1 \frac{d^2 u_1}{d\xi^2} + c \frac{du_1}{d\xi} + f_1(u_1, ..., u_n) = 0 \qquad (2.6)$$

is replaced by

$$\begin{aligned} \frac{du_1}{d\xi} &= U_1, \\ \frac{dU_1}{d\xi} &= -\frac{c}{d_1} U_1 - \frac{1}{d_1} f_1(u_1, ..., u_n). \end{aligned} \qquad (2.7)$$

So if there are n equations in the system (2.4), there will be $2n$ variables in the associated system of the first-order equations. The dynamical system associated with it is $2n$-dimensional and is given by

$$\frac{du_1}{d\xi} = U_1,$$

$$\frac{du_2}{d\xi} = U_2,$$

$$\cdots$$

$$\frac{du_n}{d\xi} = U_n,$$

$$\frac{dU_1}{d\xi} = -\frac{c}{d_1}U_1 - \frac{1}{d_1}f_1(u_1, ..., u_n)$$

$$\frac{dU_2}{d\xi} = -\frac{c}{d_2}U_2 - \frac{1}{d_2}f_2(u_1, ..., u_n)$$

$$\cdots$$

$$\frac{dU_n}{d\xi} = -\frac{c}{d_n}U_n - \frac{1}{d_n}f_n(u_1, ..., u_n).$$

(2.8)

We note that there is more than one way to turn the equation (2.6) into a system of two equations of the first order, but all of the resultant systems are equivalent, so the outcome of the existence analysis holds for all of these.

Let us say that the system (2.4) is partly parabolic, for example, $d_1 = 0$, then the first equation in (2.4) will be simply rewritten as one equation

$$\frac{du_1}{d\xi} = -\frac{1}{c}f_1(u_1, ..., u_n),$$

and so the resultant dynamical system has $2n - 1$ or less dimensions. It is easier to work with lower dimensional systems, therefore often when one of the diffusion coefficients is much smaller than the other diffusion coefficients and the other parameters in the system, the system can be "simplified" by setting that small parameter to be zero. We put quotation marks around the word "simplified" because although the existence of the waves proof may be simpler, one has to justify that the results in the "simplified" system extend to the system with the actual, nonzero, although small diffusion coefficient. Generally speaking, it is not always the case that the limit of the system with a nonzero, but small, diffusion coefficient d_1 as $d_1 \to 0$ is continuous in d_1. This type of a limit is called singular and the system that contains the equation with a non-zero small d_1 is called a singular perturbation of the system where d_1 is replaced by zero. There exists is a method of analyzing singularly perturbed systems. It is called the Geometric Singular Perturbation Theory [58, 69, 95]. Geometric Singular Perturbation Theory provides tools for checking if the existence of the solutions in the limiting system implies the existence of solutions in the system with a nonzero, but sufficiently small parameter. The situation with the stability analysis is complicated too. There are major differences in the stability analysis of the waves in parabolic reaction-diffusion systems and partly parabolic systems. We discuss these differences in Section 4.4.

As in the case of the scalar reaction-diffusion equation, the constant equilibria and the traveling wave solutions in the scalar equation have their counterparts in the associated system of the first-order equation. Assume that the latter has k equations, then it is a dynamical system over k-dimensional phase space: the expressions on the right side of the equality sign in the system (2.8) define a direction. In this phase space a constant solution $(u_1, ..., u_n) = (a_1, ..., a_n)$ to (2.8) corresponds to a point (an equilibrium point) A in the k-dimensional phase space. Since the system (2.8) of the first-order equations was formed using the approach given in (2.7), then all of the participating derivatives U_k of the said equilibrium point are equal to zero. A pulse which is an n-component vector function that satisfies the system (2.4) and which has that has $(u_1, ..., u_n) = (a_1, ..., a_n)$ as a rest state, corresponds to a homoclinic orbit that as $\xi \to \pm\infty$ converges to A. A front corresponds to a heteroclinic orbit. A periodic solution corresponds to a closed orbit. It is impossible to visualize these objects in the space of the dimension four or higher, but the meanings of the terms are the same as in the case of a two-dimensional dynamical system.

2.2 EXAMPLES OF REACTION-DIFFUSION SYSTEMS

2.2.1 FitzHugh-Nagumo system

In 1939 Alan Lloyd Hodgkin and Andrew Fielding Huxley published a paper [46] about their experiments on the squid giant axon which is typically about 0.5 mm in diameter. During their experiments they came up with a way to record the potential difference across the membrane of the axon. Hodgkin and Huxley resumed working in that direction after the World War II, in 1946. Their experimental and analytical work resulted in a series of papers [47, 49–52] where they explained the mechanism of the action potential propagation in the squid's giant axon and formulated a mathematical model for it in [48]. The story of their collaboration is described in the paper [90]. The paper includes description of their experiments on the squid giant axons and the explanation of the analysis that led to the formulation of the Hodgkin-Huxley model.

In 1963, Hodgkin and Huxley received the Nobel Prize in Physiology or Medicine "for their discoveries concerning the ionic mechanisms involved in excitation and inhibition in the peripheral and central portions of the nerve cell membrane" [45]. We refer the reader to the Award Ceremony speech [39] which explains with a good level of detail both the experiments and the explanations.

As we mentioned in Section 1.5.2, the Hodgkin-Huxley model consists of several nonlinear coupled ordinary differential equations that describe the evolution in time of the voltage and the variables associated with the activation of the sodium channel and the activation and inactivation of the potassium channel. Since the system was difficult to analyze, it was important to formulate models that captured the important features of the system but were simpler. One such model was offered by Richard FitzHugh[1] [27] in 1961. This model contained two unknown functions of time, the voltage and a variable that was responsible for the activation and deactivation mechanism.

FitzHugh's model as well as the original Hodgkin-Huxley model used the total membrane current per unit area as one of the participating parameters. Nagumo et al. in a paper published in 1962 [80] further modified the Hodgkin-Huxley and FitzHugh's models. In particular, it was suggested in [80] to use the distance along the axon as a spatial variable and to rewrite the current in terms of the second derivative of the voltage with respect to this spatial variable. Using the space variable in addition to the time resulted in better understanding of the mechanism of the propagation of the electric signal along the axon. This modification of the FitzHugh's model became known as the FitzHugh-Nagumo model. The model reads:

$$\frac{1}{c}u_t = du_{xx} + w + u - \frac{u^3}{3},$$
$$cw_t = a - bw - u. \tag{2.9}$$

This model consists of two coupled partial differential equations that describe the evolution of two quantities instead of four in the Hodgkin-Huxley system. The variable u represents membrane potential or the

---

[1]Richard FitzHugh (1922–2007) was an American biophysicist known for his fundamental contributions to mathematical neuroscience and cardio-dynamics, in particular.

voltage along the axon of a single neuron. It plays the role of the membrane voltage and the sodium activation combined. It is of a critical importance that $u - \frac{u^3}{3}$ be a cubic function that simulates action potential. Indeed, action potential is a fluctuation of the membrane potential which occurs as a rapid upward spike in the potential followed by a rapid fall and then recovery. Cubic functions capture this behavior. The variable w is called the recovery variable. It plays the role of both the sodium and the potassium inactivation together and is responsible for the mechanism of the resetting of the membrane. The constants a, b, c, and d are such that $0 < b < 1$, $1 - \frac{2}{3}b < a < 1$, $c^2 < b$, $d > 0$. We omit the description of their physical meanings. Interested readers can find the details in the original paper [80].

The system (2.9) is partly parabolic. Only the first-equation in the system has the diffusion term as the quantity u is allowed to diffuse. The second equation is a first-order partial differential equation, since the quantity w does not diffuse.

The system (2.9) has been extensively studied. Many of the results are formulated for a rescaled version of (2.9),

$$
\begin{aligned}
U_t &= DU_{xx} - W + U(U - \alpha)(1 - U), \\
W_t &= \epsilon(U - \gamma W).
\end{aligned}
\tag{2.10}
$$

The system (2.10) is obtained from (2.9) by a rescaling of the participating variables (equation (1.3) defines what is meant by a rescaling in the case of one dependent variable). This version of the system is convenient because the equilibrium important for the mechanism of the action potential propagation in the axon is now the point $(U, W) = (0, 0)$. The system (2.10) supports pulses that have $(U, W) = (0, 0)$ as their asymptotic limit and fronts where this point is one of the rest states. It also supports periodic waves. The literature devoted to the investigation of the waves in this equation is extensive (see for example [15, 17, 43, 103]). Many fundamental results are obtained for the case when ϵ is very small. In the presence of a small parameter, the system (2.10) has a multi-scale structure. Certain phenomena occur on a very slow time scale and certain are much faster.

The FitzHugh-Nagumo model was important not only because it led to better understanding of the fundamental processes of nerve conductance, but also because it facilitated the development of new mathematical techniques, especially in relation to the stability of the pulses [57].

2.2.2 Population models

Reaction-diffusion systems that model interactions between populations of predators and prey have been studied analytically and numerically. The interest toward new and classical predator-prey systems is current. Usually these systems are set in one or two space dimensions. Because of their nonlinear nature, population models in numerical simulations display a wide variety of spatiotemporal patterns that are not always well-understood. In this section we introduce some of the well-known predator-prey reaction-diffusion systems.

Many population models consist of ordinary differential equations which describe the time-evolution of a population. One of the first models that captures the predator-prey dynamics is the system

$$
\begin{aligned}
u_t &= \alpha u - \beta u v, \\
v_t &= -mv + \delta u v.
\end{aligned}
\tag{2.11}
$$

This model is called the Lotka-Volterra model. It was published in 1925 by Lotka[2] [73] in the context of chemical kinetics and independently developed by Volterra[3] [99] as a competition model. It is considered to be a classical predator-prey model. Here $u = u(t)$ plays a role of the prey population and $v = v(t)$ is the predator population. The constant parameters α, β, m, and δ are assumed to be positive. The equation for the prey contains the term αu for the growth of the prey population when there are no predators. Without predators, the population of the prey increases exponentially. When there are predators, the term $-\beta u v$ is responsible for the decrease of the population due to the predation. One can think of this term as a quantity proportional to the number of encounters of the predator and prey. A similar term $\delta u v$ contributes to the growth of the population of the predator. The $-mv$ describes the rate at which the predator will die out if there is no prey available. When there is no prey, the population of predators decreases exponentially fast.

The dependence of the populations on the space variable may be added to this equation through the diffusion terms for one or both of

---

[2]Alfred James Lotka (1880–1949) was an American mathematician, physical chemist, statistician, and biophysicyst famous for his work in population dynamics

[3]Vito Volterra (1860–1940) was an Italian mathematician and physicist, known for his fundamental contributions to functional analysis, mathematical biology, and theory of integral equations.

the involved quantities,

$$u_t = d_u u_{xx} + \alpha u - \beta uv,$$
$$v_t = d_v u_{xx} - mv + \delta uv.$$

$$(2.12)$$

The constant equilibria for the systems (2.11) and (2.12) are the same. It is the trivial equilibrium where there is no prey and no predator $(u, v) = (0, 0)$ and the coexistence equilibrium $(u, v) = (m/\delta, \alpha/\beta)$.

Very often it is useful to take into account that populations of prey do not grow without bounds, even in absence of predators. The growth is restricted by the carrying capacity of the environment. Therefore, the term αu may be typically replaced by $\alpha u \left(1 - \frac{u}{K}\right)$,

$$u_t = d_u u_{xx} + \alpha u \left(1 - \frac{u}{K}\right) - \beta uv,$$
$$v_t = d_v u_{xx} - mv + \delta uv.$$

$$(2.13)$$

It implies that the population grows as long as $u < K$, where K is a positive constant representing the carrying capacity, and decreases when $u > K$. Introducing carrying capacity changes the constant equilibria of the system. There is still the trivial equilibrium $(0, 0)$, but the coexistence equilibrium now is found by solving the algebraic system

$$\alpha \left(1 - \frac{u}{K}\right) - \beta v = 0, \quad -m + \delta u = 0,$$

$$(2.14)$$

and, thus, the equilibrium is

$$\left(\frac{m}{\delta}, \frac{\delta K - m}{\delta K \beta}\right).$$

$$(2.15)$$

The following system is a generalization of the diffusive Lotka-Volterra system (2.13). It incorporates different ways the predator responds to the variations in the density of the population of the prey:

$$u_t = d_u u_{xx} + \alpha u \left(1 - \frac{u}{K}\right) - \beta f(u)v,$$
$$v_t = d_v v_{xx} - mv + \delta f(u)v.$$

$$(2.16)$$

Here $u = u(t, x)$ represents the density of a prey population and $v = v(t, x)$ represents the density of the population predators, while ϵ_u, ϵ_v, α, K, δ and m are positive parameters.

The first equation in (2.16) is the equation that models the evolution of the prey. It incorporates the diffusion at the rate ϵ_u and the intrinsic growth of the prey at rate α. The intrinsic growth is constrained by the carrying capacity K of the environment for the prey. The nonlinear term $-\beta f(u)v$ describes the rate at which the population of the prey is decreased due to predation.

The second equation models the evolution of the predator. It includes the diffusion term with the rate ϵ_v, the natural death rate of the predator described by m, and as in the first equation, a term proportional to $f(u)v$ but with a positive sign. The term $\delta f(u)v$ in this equation measures the rate at which the biomass of the prey is converted to the biomass of the predator, in other words, it captures how the consumption of the prey contributes to the growth of the population of the predator.

Depending on the form of the function f, the system (2.16) describes different predator-prey population interactions. Holling[4] in [53] introduced three main types of the functional response of the predator to the density of the prey population.

A case with linear function $f(u) = u$ is called the Holling type I functional response. It corresponds to assumption that the predation rate depends only on the size of the population of the prey. An example of a system the data for which fits the type I functional response is described in [7]: the functional response of the brown lemmings foraging in arctic tundra was shown to be linear. This is a predator-prey system with the lemmings as the predator and the plants they eat as the prey. As such, it is a case of a herbivore-plant system. The reaction-diffusion system (2.16) then takes the form

$$u_t = \epsilon_u u_{xx} + \alpha u \left(1 - \frac{u}{K}\right) - \beta uv,$$
$$w_t = \epsilon_v v_{xx} + \delta uv - mv. \tag{2.17}$$

This system is related to the classical predator-prey Lotka-Volterra model [21, 99]. More precisely, the Lotka-Volterra model is the system (2.17) without the term $\frac{u}{K}$.

The Holling type II functional response is given by the nonlinear function $f(u) = \frac{u}{u+1}$. In this case the rate of the consumption of the

---

[4]Crawford Stanley Holling (1930–2019) was a Canadian ecologist and forest entomologists, known his significant contributions to ecology, environmental management, and ecological economics.

prey increases as the population of the prey increases but not without a bound. The increase is hindered, for example, by the limitations on the predation rate due to the time the predator needs to capture, kill, and consume the prey or by their appetite. For example, it was observed [101] that stink bugs are willing to eat an average of two bean beetles a day, even when there are plenty of beetles to attack. The following system with the Holling type II functional response is called Rosenzweig-MacArthur model,

$$u_t = \epsilon_u u_{xx} + \alpha u \left(1 - \frac{u}{K} \right) - \alpha \frac{u}{u+1} v,$$
$$w_t = \epsilon_v v_{xx} + \delta \frac{u}{u+1} v - mv. \tag{2.18}$$

Here the consumption of prey by a unit number of predators saturate at the value $1 = \lim_{u \to \infty} \frac{u}{u+1}$. In the model (2.18) it is assumed that the predator w does not have food sources other than u, so the same term $\frac{u}{u+1}$ regulates the increase of the population of predators and the decrease of the population of prey due to the predation. Adjustments may be made to incorporate the second source of food or the environmental protection or adaptability to the level of predation of the predator or prey species by replacing $\frac{u}{u+1}$ with $\frac{u}{u+a_i}$, for some $a_1 \neq a_2$ [2],

$$u_t = \epsilon_u u_{xx} + \alpha u \left(1 - \frac{u}{K} \right) - \alpha \frac{u}{u+a_1} v,$$
$$w_t = \epsilon_v v_{xx} + \delta \frac{u}{u+a_2} v - mv. \tag{2.19}$$

The Holling type III functional response is given by the nonlinear function $f(u) = \frac{u^2}{u^2+1}$. As in the type II functional response, the rate of the consumption of the prey increases as the population of the prey increases but stays bounded and has an asymptotic limit, but type III describes the situation where for the lower prey density the predation rate is smaller than with the type II functional response. In realistic predator-prey systems that happens when, for example, the predator switches to a different food source when the prey is not abundant like the avian predators feeding on spruce budworms [79, 14.7], or when the predator learns more effective hunting tactics as it gets exposed to more prey at the higher prey density as in the example of shrews and deer mice feeding on sawflies [53]. A system with the Holling type III functional response reads

$$
u_t = \epsilon_u u_{xx} + \alpha u \left(1 - \frac{u}{K}\right) - \alpha \frac{u^2}{u^2 + 1} v,
$$
$$
w_t = \epsilon_v v_{xx} + \delta \frac{u^2}{u^2 + 1} v - mv.
$$
(2.20)

All of these models exhibit a variety of patterns and waves. These have been studied extensively, but because so many parameters are present in these equations, there are still many regimes to be studied.

2.2.3 Belousov-Zhabotinski reaction

Research focused on oscillatory phenomena in chemical reactions started at the beginning of the 20th century through the analysis of theoretical models. In 1910 Lotka published a paper in *The Journal of Physical Chemistry*, where he derived a nonlinearly coupled system of two differential equations as a theoretical model for a chemical reaction that exhibited dumped oscillatory behavior characterized by the oscillations converging to an equilibrium state. In 1920, Lotka published another paper where he considered a different system of differential equations that modeled an autocatalytic reaction exhibiting undamped oscillations. While in the 1910 paper Lotka doubted the possibility of the existence of self-sustained oscillations in chemical reactions, in the 1920 paper he derived the system (2.11) known as Lotka-Volterra system. In his formulation, u and v represent concentrations of reacting chemicals. Lotka numerically found some periodic solutions in (2.11), more precisely, this system is a model for self-sustained oscillations. The associated system of partial differential equations that takes into account spatial dependence has the form (2.12).

Between 1920 and 1959, there were not many publications on the research related to the oscillatory reactions. However, the discovery of an oscillatory reaction by Bray in 1921 is of particular importance, because the community of researchers, both experimentalists and theorists, did not believe that chemical reactions may support sustained oscillations, which they explained by the second law of thermodynamics. Around 1950, P. Belousov[5] in his experiments observed temporal oscillations in a chemical reaction, as well as other complex structures such as traveling waves. He unsuccessfully tried to publish his results, facing rejection

---

[5]Boris Pavlovich Belousov (1893–1970) was a chemist and biophysicist from the Soviet Union who discovered the Belousov-Zhabotinsky reaction.

and harsh criticism [102]. In 1959, he managed to finally publish a brief abstract of his work. The original paper was published in full in Russian in 1981 and appeared in English in the Oscillations and traveling waves in chemical systems edited by Field and Burger in 1988 [9].

Belousov's discovery marked the beginning of active research in oscillatory chemical reactions. In 1965, the following system

$$u'(t) = k_1 cu - k_1 uv - k_0 uw,$$
$$v'(t) = k_1 cu - k_1 uv - k_2 v, \qquad (2.21)$$
$$w'(t) = k_2 v - k_3 w,$$

was derived as a mathematical model for Belousov reaction by Zhabotinsky (Zhabotinskii) and Korzukhin [67]. In (2.21) all of the constant parameters are positive.

There are several well-known systems of differential equations derived by different groups of researchers as models for the Belousov-Zhabotinsky (BZ) reaction. One of the earlier models was called the Brusselator model formulated by I. Prigogine and R. Lefever in 1968 in Brussels. A modification of the Brusselator model was called Oregonator after Oregon, where the research by Field, Kórós, and Noyes [22,23] took place. The Oregonator model reads

$$u'(t) = k_1 av - k_2 uv + k_3 au - 2k_4 u^2,$$
$$v'(t) = -k_1 av - k_2 uv + fk_5 w, \qquad (2.22)$$
$$w'(t) = k_2 bu - k_5 w,$$

where all parameters are positive. The Oregonator model has been studied for different types of oscillatory solutions that encode the temporal oscillations in the BZ chemical reaction. In order to understand the formation of spatiotemporal patterns and their qualitative and quantitative properties, many research groups studied reaction-diffusion systems associated with the corresponding chemical kinetics equations. For example, the reaction-diffusion system associated with (2.22) is the following system of partial differential equations,

$$u_t = d_u u_{xx} + k_1 av - k_2 uv + k_3 au - 2k_4 u^2,$$
$$v_t = d_v v_{xx} - k_1 av - k_2 uv + fk_5 w, \qquad (2.23)$$
$$w_t = d_w w_{xx} + k_2 bu - k_5 w,$$

where d_u, d_v, $d_w > 0$ are the diffusion constants and x represents the space variable.

Since these systems are complex and involve many equations, several attempts have been made to simplify these equations, while still preserving their important properties. For example, the following scalar equation

$$u_t = d_u u_{xx} + k_3 au - 2k_4 u^2 \qquad (2.24)$$

was suggested for studying traveling waves [40]. In (2.24), $d > 0$ is the diffusion constant and x is the space variable. Note that the equation (2.24) is a Fisher-KPP equation. Indeed, the equation (2.24) looks like the equation (1.27) if the equation (2.24) is divided by $2k_4$ and then t is replaced with $\tau = 2k_4 t$ and new parameters $d = \frac{d_u}{2k_4}$ and $m = \frac{k_3 a}{2k_4}$ are introduced.

On the other hand, in [78], Murray presented a system alternative to the diffusive version of the Oregonator model given in (2.23), which was obtained by considering modified equations of the chemical kinetics:

$$
\begin{aligned}
u_t &= d_u u_{xx} + k_1 av - k_2 uv + k_3 au - 2k_4 u^2, \\
v_t &= d_v v_{xx} - k_1 av - k_2 uv
\end{aligned}
\qquad (2.25)
$$

and used it to study traveling front solutions.

While the topic of traveling fronts in equations that model the BZ reaction have been pursued by various researchers, solutions of oscillatory nature were of a particular interest. Tyson and Fife in [96] introduced a new scaling and rewrote (2.23) as

$$
\begin{aligned}
u_t &= \epsilon u_{xx} + \frac{1}{\epsilon}(qv - uv + u - u^2), \\
v_t &= \gamma \epsilon v_{xx} + \frac{1}{\eta}(-qv - uv + fw), \\
w_t &= \delta \epsilon w_{xx} + u - w.
\end{aligned}
\qquad (2.26)
$$

They then considered the limit of the system (2.26) when $\delta = 0$ and $\eta = 0$ and formulated the system

$$
\begin{aligned}
u_t &= \epsilon u_{xx} + \frac{1}{\epsilon}\left(u - u^2 - \frac{fw(u-q)}{u+q}\right), \\
w_t &= u - w
\end{aligned}
\qquad (2.27)
$$

and showed that the system (2.27) supports spatiotemporal periodic solutions. In two-dimensional physical space, these patterns are target patterns, in other words, rings propagating out from a central point.

Although many results have been published about models for the BZ reaction, traveling fronts and oscillatory patterns in these models are still very much active research topics.

2.2.4 Spread of infection disease

In this section we give examples of systems of differential equations that model disease dynamics in a single population. In the presence of a certain contagious disease, the population is partitioned into two groups. One group contains the members of the population that are susceptible to the infection and the second group includes the members that carry the infection or the infected. In the process of interaction of the infected members of the population with the members that are susceptible, the disease transmission occurs within the "well-mixed" population. The following system of ordinary differential equations

$$\frac{\mathrm{d}S}{\mathrm{d}t} = -\beta S(t)I(t) + \gamma I(t),$$
$$\frac{\mathrm{d}I}{\mathrm{d}t} = \beta S(t)I(t) - \gamma I(t), \tag{2.28}$$

is a basic local model, called as SIS model, for the transmission of an infection in a "well-mixed" population [72, 91]. In (2.28), $S(t) \geq 0$ is the fraction of the population at time t that is susceptible to the disease and $I(t) \geq 0$ is the fraction of the infected members of the population. Parameter $\beta > 0$ measures the rate at which members of the population are infected and $\gamma > 0$ measures the rate of the recovery of infected members.

If it is assumed that members of a population move through a spatial domain, then diffusion terms (terms proportional to the second spatial derivatives of the participating quantities) may be added to the system (2.28) to obtain a system of partial differential equations

$$S_t = D_1 S_{xx} - \beta SI + \gamma I,$$
$$I_t = D_2 I_{xx} + \beta SI - \gamma I, \tag{2.29}$$

where the diffusion coefficients D_1 and D_2 are assumed to be positive constants.

The system (2.29) is also a basic model for a spread of an infectious disease in space (see [8] and references therein).

Another example of a temporal evolution of an infection is a system of differential equations known as a simplified version of Kermack and McKendrick model or the SIR model [12, 72],

$$
\begin{aligned}
\frac{\mathrm{d}S}{\mathrm{d}t} &= -\beta S(t)I(t) + \alpha R(t), \\
\frac{\mathrm{d}I}{\mathrm{d}t} &= \beta S(t)I(t) - \gamma I(t), \\
\frac{\mathrm{d}R}{\mathrm{d}t} &= \gamma I(t) - \alpha R(t).
\end{aligned}
\tag{2.30}
$$

In this model, it is assumed that the immunity received from the disease is not permanent. In the system (2.30), $S(t)$ is the fraction of the total population that is susceptible, $I(t)$ is the fraction of the population that is infected at time t, and $R(t)$ is the fraction of the population that consists of the people recovered from the infection. Parameter $\beta > 0$ measures the rate at which the infection is transmitted within the population, $\alpha > 0$ measures the rate at which the infected recover, and $\gamma > 0$ measures the rate at which the immunity of recovered members of the population wears out and they become susceptible to the disease again.

To incorporated spatial spread of the disease, diffusion terms may be added to the system (2.30), resulting in a nonlinearly coupled system of partial differential equations

$$
\begin{aligned}
S_t &= D_1 S_{xx} - \beta SI + \alpha R, \\
I_t &= D_2 I_{xx} + \beta SI - \gamma I, \\
R_t &= D_3 R_{xx} + \gamma I - \alpha R,
\end{aligned}
\tag{2.31}
$$

where the rate of dispersal of susceptible, infected and recovered members of the population incorporated in the model through the positive diffusion coefficient D_1, D_2, and D_3. We point out that if the infected people cannot move due to the disease, then one would set $D_2 = 0$ or take D_2 to be much smaller than the other diffusion coefficients.

Predicting patterns that may arise in the diffusive models such as (2.29) and (2.31) may be important to understand, control, and mitigate the effects of the spatio-temporal disease spread.

2.2.5 The high Lewis number combustion model

Systems of reaction-diffusion equations are very often used in combustion theory. The example that we discuss here is called the high Lewis number combustion model. The model consists of two partial differential equations. One equation describes the evolution of the temperature of the fuel which we denote u. The other equation is for the evolution of the concentration of the fuel in the reaction zone, which we denote y. The explanation of the deduction of the model may be found in [71] and the references therein. Here we introduce the model and explain what the terms in it mean. We point out that u and y are non-dimensionalized, which means that a scaling was applied to the underlying physical laws that effectively canceled the units of measurement for all of the participating quantities. The following assumptions on the combustion process are made that lead to a simpler model: the fuel is premixed and there is no heat loss in the system. We formulate the system in one spatial variable $x \in \mathbb{R}$, so the temperature u and the concentration of the fuel y are functions of the spatial variable x and time t. Using a one-dimensional spatial variable is justified for the case when the system describes burning of a fuel arranged in one-dimensional configuration, for example, in a long thin pipe. In addition, it serves as a reasonable approximation for a system describing propagation of a planar combustion front.

The system describing evolution of the temperature u and concentration of the fuel y reads

$$\begin{aligned} u_t &= u_{xx} + y\Omega(u), \\ y_t &= \epsilon y_{xx} - \beta y\Omega(u). \end{aligned} \tag{2.32}$$

In the reaction term, the function Ω has the form

$$\Omega(u) = \begin{cases} e^{-1/u}, & \text{for } u > 0, \\ 0, & \text{otherwise.} \end{cases} \tag{2.33}$$

This kind of reaction term is called an Arrhenius law without ignition cut-off. The temperature $u = 0$ represents a background temperature at which the reaction does not yet take place. The presence of the positive ignition cut-off means that the combustion process starts at a temperature strictly above the background temperature.

The system (2.32) has two parameters. One is the exothermicity parameter $\beta > 0$ which is the ratio of the activation energy to the heat of the reaction. In other words, $\beta > 0$ shows how much fuel should be burned to achieve a unit increase in temperature. The other parameter ϵ is the reciprocal of the Lewis number $\epsilon = 1/\text{Le} > 0$. The Lewis number Le is a physical constant that represents the ratio of the heat diffusivity to the fuel diffusivity.

The model is called the high Lewis number combustion model when Le is a very large number, and so ϵ is very small. Dense liquid fuels or fuels with a liquid phase (which means they become liquid at high temperatures) have a very large Lewis number, so $0 < \epsilon \ll 1$. An example of liquid fuel combustion is related to burning of toxic wastes.

The system (2.32) in a moving frame reads

$$u_t = u_{\xi\xi} + c\,u_\xi + y\,\Omega(u),$$
$$y_t = \epsilon y_{\xi\xi} + c\,y_\xi - \beta\,y\,\Omega(u). \tag{2.34}$$

We are looking for traveling fronts in this system, so we obtain the traveling wave equations by setting the time derivatives equal to zero

$$u'' + c\,u' + y\,\Omega(u) = 0,$$
$$\epsilon y'' + c\,y' - \beta\,y\,\Omega(u) = 0, \tag{2.35}$$

where the derivative is with respect to ξ.

There is a continuum of equilibrium points for the system (2.32) obtained by solving the equation $y\Omega(u) = 0$. The equilibria are $\{(u, y) : y = 0\}$ and $\{(u, y) : u = 0\}$. We note that based on the physical meaning, it makes sense to consider only nonnegative u and y. The maximal physical value of the concentration of the fuel is $y = 1$, because of the scaling applied when the system was deduced. The reaction is assumed to start at $u = 0$. When $y = 1$, to obtain an equilibrium we should take $u = 0$ in $y\Omega(u) = 0$. This equilibrium $(u, y) = (0, 1)$ represents the state where all of the fuel is present and the reaction is about to start. Another state clearly identifiable at the physical level is when all of the fuel is burned: $y = 0$. From the mathematical point of view, the temperature at this state may be any positive constant, but let us take a closer look at the traveling equations (2.35). If we multiply the first equation in (2.35) by β and add this equation to the second equation, we get

$$\beta u'' + \beta c\,u' + \epsilon y'' + c\,y' = 0. \tag{2.36}$$

All of the terms on this equation are derivatives, so we can integrate it,

$$\beta u' + \beta c\,u + \epsilon y' + c\,y = C. \tag{2.37}$$

Here C is the constant of integration. This equation tells us that the quantity $\beta u' + \beta c\,u + \epsilon y' + c\,y$ is conserved. We assume that we are searching for the waves that have $(u, y) = (0, 1)$ as one of its rest states, say, at $+\infty$, so the combustion front propagates toward the state where all of the fuel is still present. We then pass to a limit $\xi \to +\infty$ in (2.37) and see that the integration constant C must be equal to the speed of the front c which we do not know yet, so

$$\beta u' + \beta c u + \epsilon y' + c y = c. \tag{2.38}$$

Now assume that the front at $-\infty$ rests at $y = 0$, in other words, a state where all of the fuel is completely burned is left behind the front. We then take a limit $\xi \to -\infty$ in (2.38) and obtain an equation that defines the value of u that corresponds to $y = 0$, which is $\beta c u = c$, which leads to $u = 1/\beta$. Therefore, the equilibrium at the $-\infty$ limit is $(u, y) = (1/\beta, 0)$. The value $u = 1/\beta$ may be interpreted as the maximal value of the temperature that may be reached in the system. An important solution for the model (2.32) is a combustion front that asymptotically connects these two equilibria, or more precisely, a solutions of (2.35) that satisfies the following boundary-like conditions:

$$
\begin{aligned}
(u_0, y_0) &\to (0, 1) \text{ as } \xi \to \infty \text{ and} \\
(u_0, y_0) &\to (u_B, 0) \text{ as } \xi \to -\infty, \quad u_B = 1/\beta.
\end{aligned}
\tag{2.39}
$$

The model (2.32) is a system of two reaction-diffusion equations when $\epsilon > 0$. A related system is obtained from (2.32) by setting $\epsilon = 0$,

$$
\begin{aligned}
u_t &= u_{xx} + y\Omega(u), \\
y_t &= -\beta y \Omega(u).
\end{aligned}
\tag{2.40}
$$

With an abuse of notation we write the system (2.40) in the following form with an understanding that the reaction occurs only for $u > 0$,

$$
\begin{aligned}
u_t &= u_{xx} + y\,e^{-1/u}, \\
y_t &= -\beta\,y\,e^{-1/u}.
\end{aligned}
\tag{2.41}
$$

The system in a moving frame reads

$$
\begin{aligned}
u_t &= u_{\xi\xi} + c\,u_\xi + y\,e^{-1/u}, \\
y_t &= c\,y_\xi - \beta\,y\,e^{-1/u}.
\end{aligned}
\tag{2.42}
$$

The traveling wave solutions such as fronts then satisfy the system

$$
\begin{aligned}
u'' + c\,u' + y\,e^{-1/u} &= 0, \\
c\,y' - \beta\,y\,e^{-1/u} &= 0,
\end{aligned}
\tag{2.43}
$$

where the prime denotes an differentiation with respect to ξ. Similarly to the case of nonzero ϵ, for (2.40) an important solution is a combustion front that satisfies the equation (2.43) together with boundary conditions (2.39). The front is a solution of the system (2.43) and the components of the front satisfy a conserved quantity, which is the equation (2.38) with $\epsilon = 0$,

$$\beta u' + \beta c u + c y = c. \tag{2.44}$$

The first equation in (2.42) is an equation of the second order. To simplify the system we replace the first equation in (2.43) with the equation (2.44). Rewriting both equations in the normal form, i.e., solving the equations for the highest derivatives, we then obtain

$$u' = \frac{c}{\beta}(1 - y - \beta u),$$
$$y' = \frac{\beta}{c} y e^{-1/u}. \tag{2.45}$$

With our definition of ϵ as the reciprocal of the Lewis number, taking $\epsilon = 0$ is equivalent to taking $\text{Le} = \infty$. The latter is a characteristic of the solid fuels, therefore, the system (2.41) describes the combustion processes that have only the solid phase, with no gaseous products present. Typical examples of solid fuels are coal or wood, and also charcoal, peat, coal, biomass, biomass-coal, and bagasse which is the fibrous material remaining after the juice is pressed from sugar cane. With solid fuels, the equation for the temperature is a reaction-diffusion equation because the temperature diffuses, but the equation for the amount of fuel lacks the diffusion term since the solid fuels do not diffuse. The second equation then is not a reaction-diffusion equation. This is an example of a partly parabolic system discussed in Section 2.1.

The systems (2.32) and (2.40) are models of realistic combustion processes that capture propagation of combustion fronts, therefore each of these systems is of interest. In addition, the transition between the cases of zero and nonzero ϵ is also important to understand because sometimes liquefaction of the solid fuel occurs in the reaction zone, thus leading to a non-zero value of $\epsilon = 1/\text{Le}$. Another argument is that the system (2.42) originated in physics as an approximation of the high Lewis number model: the very small ϵ in the system (2.32) is replaced by $\epsilon = 0$ (see [71] and the references therein). It is not obvious whether the systems with zero and nonzero but small ϵ have the same properties, since the limit $\epsilon \to 0$ is singular. By "singular" here we mean that when applying the limit $\epsilon \to 0$ to system (2.35) to obtain (2.43), one goes from a fourth order system of differential equations to a third order one, thus reducing the order by 1. If the purpose of setting $\epsilon = 0$ is to simplify the model, one has to justify that the system (2.40) is indeed an approximation of the system (2.32). The study of the relation between of zero and nonzero ϵ has been already conducted and it was shown that for sufficiently small ϵ, the combustion front solution in the system (2.32) is an ϵ-order perturbation of the combustion front solution in the system (2.40).

In Section 3.4 we present a proof of the existence of a solution of (2.45) that satisfy boundary conditions (2.39), and thus, is a traveling front for the original system (2.40).

CHAPTER 3

Existence of fronts, pulses, and wavetrains

3.1 TRAVELING WAVES AS ORBITS IN THE ASSOCIATED DYNAMICAL SYSTEMS

In this section we discuss dynamical system approach to the investigation of traveling waves in the reaction-diffusion equation (1.4), which we rewrite here for the convenience of the reader,

$$u_t = u_{xx} + f(u). \tag{3.1}$$

The goal of this chapter is to bring examples of techniques, analytical and numerical, that can prove or confirm the existence of traveling wave solutions and/or allow to visualize the waves as functions of the traveling wave coordinate $\xi = x - ct$ or as parametric curves in the phase space which is the space with axes (u, u_ξ).

As we discussed before, traveling waves are solutions of the traveling wave equation (1.12), which reads

$$0 = u'' + cu' + f(u), \tag{3.2}$$

where the derivative is with respect of the moving coordinate $\xi = x - ct$. Recall that we do not yet know the value of the speed parameter c. The equation may be accompanied by boundary conditions (1.23),

$$\lim_{\xi \to -\infty} u(\xi) = B, \quad \lim_{\xi \to +\infty} u(\xi) = A, \tag{3.3}$$

if we are looking for pulses $(A = B)$ or fronts $(A \neq B)$, or, for periodic wavetrains, by the expectation of the periodicity of the solution.

DOI: 10.1201/9781003147619-3

Exact solutions of (3.2) may be found in some particular situations. That is the case with the KdV equation (1.49) discussed in Section 1.6.3 for which the traveling wave equation (1.50) was shown to have families of traveling wave solutions described by the formula (1.62). Sometimes, it may be possible to find a pulse or a front by fitting a hyperbolic function for a solution [75, 76]. The hyperbolic tangent function is used to find monotone fronts and the hyperbolic secant function is used to find a pulse.

In general, we do not expect that a formula for a traveling wave can be found. Instead, we try to prove the existence of the traveling waves, without trying to solve the equation. The approach that we describe below is based on interpreting the ordinary differential equation (3.2) as a dynamical system. A dynamical system is associated with a system of the first-order ordinary differential equations and its vector field which is interpreted as a mapping of a point A_0 from the phase space of the system and a value of the independent variable, say ξ into a point $A(\xi)$ from the phase space. Geometrically, a dynamical system associated with a system of the first-order differential equations describes the movement in the phase space along the solution curves [84].

It is easy to see that the equation (3.2) is equivalent to the system

$$
\begin{aligned}
u' &= v, \\
v' &= -cv - f(u),
\end{aligned}
\tag{3.4}
$$

where the derivative is taken with respect to the moving coordinate ξ. The term equivalent here means that if we have a solution to the system (3.4) then we know the solution to the equation (3.2) and vice versa. A solution of (3.4) consists of a pair of functions $(u, v) = (u(\xi), v(\xi))$ that satisfy the system (3.4), so if we have a solution of (3.4) then its u-component solves the equation (3.2). On the other hand, if we have a solution u of (3.2), then we can also calculate $v = u'$ in order to get the solution of (3.4).

The solutions of the system (3.4) may be interpreted as parametric curves in the (u, v)-plane, with the parameter being ξ. These curves are called trajectories. The right hand side of the system (3.4) at each point of (u, v)-plane provides the slope of the vector tangent to the solution curve that passes through the point. Therefore the system (3.4) generates a vector field over the (u, v)-plane that the solutions of the system must follow.

Existence of traveling wave solutions in (3.2) can be shown by using nonlinear dynamical systems methods through establishing the existence of the associated orbits in (3.4). Rather than trying to solve the nonlinear equation (3.2), one can analyze the vector field of the system (3.4) and show the existence of special solutions such as traveling waves and characterize their geometric properties, which can be used in understanding the stability properties of such waves using analytical and numerical methods. The textbook Nonlinear Dynamics and Chaos: With Applications to Physics, Chemistry, and Engineering by S. Strogatz [93] is an excellent entry point to the subject of nonlinear dynamics. A textbook designed for senior undergraduates is Differential Dynamical Systems by J. Meiss [77]. There are many more excellent books on nonlinear dynamics.

There is a correspondence between solutions of (3.2) and the trajectories of (3.4). For example, let constant a be such that $f(a) = 0$, then a constant solution (the equilibrium) $u = u(\xi) = a$ in (3.2) corresponds in the system (3.4) to the equilibrium $(u, v) = (a, 0)$, since the derivative of a constant is zero.

A pulse $u = u_0(\xi)$ as a solution of the equation (3.2) such that

$$\lim_{\xi \to \pm\infty} u_0(\xi) = a$$

in the system (3.4) corresponds to a solution asymptotically converging to the equilibrium $(a, 0)$ at both $\pm\infty$. Such a trajectory is called a homoclinic orbit. See Figure 3.1 for the illustration.

Assume that there are two distinct equilibrium points $u = a^\pm$ in (1.12), that is, $f(a^+) = f(a^-) = 0$ and $a^- \neq a^+$. A front u_0 as a solution of the equation (3.2) such that

$$\lim_{\xi \to \pm\infty} u_0(\xi) = a^\pm$$

in the system (3.4) corresponds to a solution asymptotically approaching $(a^+, 0)$ as $\xi \to \infty$ and to the equilibrium $(a^-, 0)$ as $\xi \to -\infty$. Such a trajectory is called a heteroclinic orbit. See Figure 3.2 for the illustration.

A periodic solution of (3.2) corresponds to a closed orbit in the system (3.4). See Figure 3.3 for the illustration of a closed orbit. While it is obvious that a homoclinic or heteroclinic orbit cannot exist in the system that does not have equilibrium points, it is not so obvious about closed orbits. In fact, existence of an equilibrium point is a necessary condition for a closed orbit to exist. This fact follows from the Index theory in dynamical systems [93].

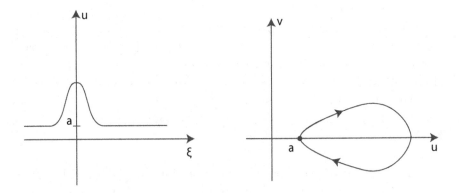

Figure 3.1 Left: A pulse solution. Right: A homoclinic orbit that corresponds to a pulse solution.

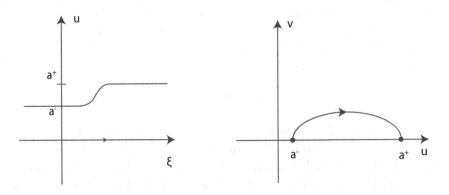

Figure 3.2 Left: A front solution. Right: A heteroclinic orbit that corresponds to a front solution.

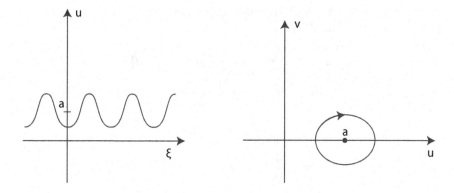

Figure 3.3 Left: A wavetrain. Right: A closed orbit that corresponds to a wavetrain.

There is more than one way to write the equation (1.12) as a system of the first-order ordinary differential equations, but all of the systems produced by the equation (1.12) are equivalent, so the choice of the system is driven by the mathematical convenience or by the intuition behind the underlying physical phenomenon.

3.2   DYNAMICAL SYSTEMS APPROACH: EQUILIBRIUM POINTS

To see whether an equation or a system of equations has a traveling wave one starts by analyzing its equilibrium points. A system without equilibrium points cannot have a traveling wave solution.

Let us say that the equation (1.12) has an equilibrium $u = a$. The system (3.4) then has an equilibrium $(u, v) = (a, 0)$. In general, there is a one-to-one correspondence between the equilibrium points of the scalar equation of the second order and the equilibrium points of the corresponding system of the first-order equations. We next take a closer look at the dynamics in the system (3.4) near the equilibrium $(a, 0)$. In a sufficiently small neighborhood of the equilibrium, one may rely on the linear approximation for the nonlinear reaction terms in (3.4). Let us denote the difference between the solution and the equilibrium as $(\tilde{u}, \tilde{v})$, in other words, let us make a substitution

$$(u(\xi), v(\xi)) = (a + \tilde{u}(\xi), 0 + \tilde{v}(\xi)) \qquad (3.5)$$

in the system (3.4). In particular, $f(u)$ becomes $f(a + \tilde{u})$, and, if we assume that $\tilde{u}$ is small, then

$$f(a + \tilde{u}) \approx f(a) + f'(a)\tilde{u} = f'(a)\tilde{u}, \tag{3.6}$$

since $f(a) = 0$. We then obtain the following system for $(\tilde{u}, \tilde{v})$,

$$\frac{d\tilde{u}}{d\xi} = \tilde{v},$$
$$\frac{d\tilde{v}}{d\xi} = -f'(a)\tilde{u} - c\tilde{v}. \tag{3.7}$$

This system is the linearization of (3.4) about $(a, 0)$. The linearized system has an equilibrium at $(0, 0)$.

Notice that the right hand side of the linear system (3.7) may be obtained by calculating the Jacobian of the vector function on the right hand side of the (3.4), more precisely, of the vector function

$$\begin{pmatrix} v \\ -cv - f(u) \end{pmatrix}, \tag{3.8}$$

which is

$$J(u, v) = \begin{pmatrix} 0 & 1 \\ -f'(u) & -c \end{pmatrix}. \tag{3.9}$$

We evaluate the Jacobian at $(u, v) = (a, 0)$,

$$J(a, 0) = \begin{pmatrix} 0 & 1 \\ -f'(a) & -c \end{pmatrix}, \tag{3.10}$$

thus obtaining the system (3.7).

The dynamics of the trajectories near the equilibrium $(0, 0)$ may be understood by looking at the eigenvalues and the eigenvectors of the Jacobian $J(a, 0)$. The eigenvalues are the values λ such that

$$\det(J(a, 0) - \lambda I) = 0, \tag{3.11}$$

where I is an identity matrix. For (3.7), they are roots of the polynomial

$$\lambda^2 + c\lambda + f'(a) = 0, \tag{3.12}$$

which are given by the formula

$$\lambda_\pm = \frac{1}{2}\left(-c \pm \sqrt{c^2 - 4f'(a)}\right). \tag{3.13}$$

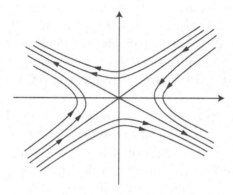

Figure 3.4 A saddle.

It is important to know if the eigenvalues $\lambda_\pm$ are real or complex. The signs of the real eigenvalues and the signs of the real parts of the complex eigenvalues are also important. These characteristics depend on the sign of $f'(a)$ and its value relative to c.

Let us assume that $f'(a) < 0$, then both $\lambda_\pm$ are real, and, moreover, $\lambda_+ > 0$ and $\lambda_- < 0$. The equilibrium $(0,0)$ is then called a saddle point. After finding the eigenvalues, we find the corresponding eigenvectors. The line spanned by the eigenvector of λ_- contains trajectories that approach the equilibrium point at $(0,0)$ at exponential rates $e^{\lambda_- t}$ at ∞. The trajectories that are in the direction spanned by the eigenvector of λ_+ move away from the equilibrium. We say that corresponding solutions converge to the equilibrium in the reverse direction of the variable ξ, at $-\infty$. To develop an intuition about the directions, one may choose to think of the independent variable ξ as the time. The other trajectories follow the vector field. Figure 3.4 contains an illustration of a generic saddle.

As long as the real parts of the eigenvalues are not zero, in other words, the eigenvalues are neither zero, nor purely imaginary, the equilibrium $(a,0)$ in (3.4) will be of the same type as the equilibrium $(0,0)$ in (3.7). The phase portrait near $(a,0)$ for (3.4) will look like a continuous deformation of the phase portrait near $(0,0)$ in the linearized system. We say that these phase portraits are topologically equivalent. A saddle is one of the types of the equilibrium points that is robust. If the linearized system has a saddle at $(0,0)$ then the original system will have a saddle at the equilibrium where the linearization was obtained. The eigenvectors of a saddle also play a big role. They are the first-order approximations for some trajectories in the nonlinear system (3.4). In other words, there are trajectories in the nonlinear system that approach the equilibrium $(a, 0)$ either forward in time (as $\xi \to \infty$) or backward in time (as $\xi \to -\infty$) for which the eigenvectors corresponding to the negative eigenvalue and the positive eigenvalue, respectively, are the tangent vectors.

Important concepts are of the stable and unstable subsets (manifolds) of an equilibrium point of first-order system. A stable subset (or a stable manifold) of an equilibrium is the set of all initial conditions such that any solution that starts in this set converges to the equilibrium point as the independent variable approaches ∞. The unstable subset (or an unstable manifold) of an equilibrium is the set of all initial conditions such that any solution that start there converges to the equilibrium as the independent variable approaches $-\infty$. For example, if the equilibrium is a saddle, then it has a one-dimensional stable and a one-dimensional unstable subsets.

If $0 < f'(a) < c^2/4$, then both $\lambda_\pm$ have the same sign as $-c$. So if $c > 0$, then both eigenvalues are negative and for the linearized system $(0, 0)$ is called a stable node. Therefore the equilibrium $(a, 0)$ is a stable node for the system (3.4). Stable and unstable nodes are also robust: the linearization nicely captures the nature of the equilibrium point. A stable node has a two-dimensional stable manifold. See Figure 3.5 for an illustration of a node.

If $f'(a) > c^2/4$, then $\lambda_\pm$ are complex conjugate numbers with real parts equal to $-c$. So if $c > 0$, then the equilibrium $(0, 0)$ for the linearized system is called a stable focus. A stable focus in the linearized system implies that the equilibrium about which the system was linearized is also a stable focus (see Figure 3.5).

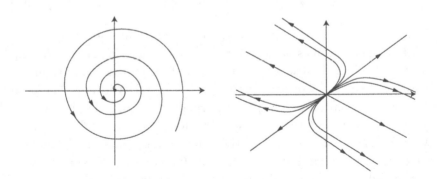

Figure 3.5 Left: An unstable node. Right: A stable focus.

Stable nodes and stable foci are also called sinks, as opposed to the unstable nodes and unstable foci which are called sources.

The tricky situations occur when the Jacobian calculated at an equilibrium has either a zero eigenvalue or purely imaginary eigenvalues. Even a smallest perturbation then may potentially change the character and the stability of the equilibrium. Such equilibria are called non-hyperbolic, as opposed to hyperbolic equilibria. Saddles, nodes and foci are hyperbolic equilibria. A study of traveling waves that are associated with a non-hyperbolic equilibrium involves a refined analysis which is outside of the scope of this book. For example, a zero eigenvalue may occur when the equilibrium is not an isolated point but is embedded in a continuum of equilibrium points which complicates the analysis. For a linear system, the set of points in the (u, v)-space such that solutions that pass through these points do not converge to that equilibrium neither forward nor backward in time form its center manifold. The sum of the dimensions of the stable, unstable and center manifolds is equal to the dimension of the phase space. For the scalar equation (3.2), this dimension is 2.

For the hyperbolic equilibria, the Stable (Unstable) Manifold theorem implies the topological equivalence of the equilibria in the linearized and the original systems. The theorem says that if a hyperbolic equilibrium at the origin in the linearized system (3.7) has a stable (an unstable) manifold of certain dimensions (the theorem applies to the systems of more than two dimensions as well), then there exists a neighborhood of the equilibrium in (3.4) where a stable (unstable) manifold of the equilibrium exists which has the same dimensions as in the linear system. The stable (unstable) manifold in the linear system is tangent to the stable (unstable) manifold in the nonlinear system (3.4).

Let us consider a more general system of differential equations of the first order,

$$
\begin{aligned}
\frac{du}{d\xi} &= f(u, v), \\
\frac{dv}{d\xi} &= g(u, v).
\end{aligned}
\tag{3.14}
$$

Assume that the system (3.14) has an equilibrium (a, b), which is found by solving a system of algebraic equations

$$
\begin{aligned}
f(u, v) &= 0, \\
g(u, v) &= 0.
\end{aligned}
\tag{3.15}
$$

To study the dynamics near the equilibrium, one linearizes the system (3.14) about the equilibrium (a, b),

$$
\begin{aligned}
\frac{d\tilde{u}}{d\xi} &= \frac{\partial f}{\partial u}(a, b)\tilde{u} + \frac{\partial f}{\partial v}(a, b)\tilde{v}, \\
\frac{d\tilde{v}}{d\xi} &= \frac{\partial g}{\partial u}(a, b)\tilde{u} + \frac{\partial g}{\partial v}(a, b)\tilde{v}.
\end{aligned}
\tag{3.16}
$$

Notice that the coefficients of $\tilde{u}$ and $\tilde{v}$ in the right hand side of (3.16) form the Jacobian $J[f, g](u, v)$ evaluated at $(u, v) = (a, b)$. The concepts described above related to the eigenvalues and the corresponding eigenvectors apply in this case as well.

3.3 EXISTENCE OF FRONTS IN FISHER-KPP EQUATION: TRAPPING REGION TECHNIQUE

In this section we will look for traveling front solutions of the Fisher-KPP equation

$$
u_t = u_{xx} + u(1 - u).
\tag{3.17}
$$

This equation has two equilibrium points $u = 0$ and $u = 1$. The traveling fronts are stationary solutions of one variable $u = u(\xi)$, where $\xi = x - ct$, $c > 0$, of the equation

$$
u_t = u_{\xi\xi} + cu_\xi + u(1 - u),
\tag{3.18}
$$

that have the equilibrium points $u = 0$ and $u = 1$ as its rest states. Therefore we seek fronts as solutions of the associated ordinary differential equation

$$
u'' + cu' + u(1 - u) = 0,
\tag{3.19}
$$

where the derivative is taken with respect to ξ, that satisfy the conditions

$$
\begin{aligned}
\lim_{\xi \to \infty} u(\xi) &= 0, \\
\lim_{\xi \to -\infty} u(\xi) &= 1.
\end{aligned}
\tag{3.20}
$$

The second-order ordinary differential equation (3.19) can be written as a system of two first-order equations

$$
\begin{aligned}
\frac{du}{d\xi} &= v, \\
\frac{dv}{d\xi} &= -cv - u(1 - u).
\end{aligned}
\tag{3.21}
$$

This system defines a vector field in the (u, v)-space. The equilibrium points for the system are found by setting the derivatives equal to zero,

$$0 = v,$$
$$0 = -cv - u(1 - u).$$

It is easy to see that there are two equilibrium points $(u, v) = (1, 0)$ and $(u, v) = (0, 0)$. The equilibria $(u, v) = (0, 0)$ and $(u, v) = (1, 0)$ of (3.21) correspond respectively to the constant states $u = 0$ and $u = 1$ of (3.19). Assume that we look for the fronts that leave $u = 1$ behind and move toward $u = 0$. We are then interested in establishing the existence of heteroclinic orbits of (3.21) that asymptotically connect equilibria $(1, 0)$ and $(0, 0)$. More precisely, we assume that the orbit satisfy the limits:

$$\lim_{\xi \to \infty} (u(\xi), v(\xi)) = (0, 0),$$
$$\lim_{\xi \to -\infty} (u(\xi), v(\xi)) = (1, 0).$$
(3.22)

In order to determine the stability properties of equilibria of the system (3.21), we linearize the system about the equilibrium points. The linearization is obtained using the Jacobian

$$J(u, v) = \begin{pmatrix} \dfrac{\partial v}{\partial u} & \dfrac{\partial v}{\partial v} \\ \dfrac{\partial(-cv - u(1-u))}{\partial u} & \dfrac{\partial(-cv - u(1-u))}{\partial v} \end{pmatrix} = \begin{pmatrix} 0 & 1 \\ -1 + 2u & -c \end{pmatrix}$$
(3.23)

of the vector function

$$\begin{pmatrix} v \\ -cv - u(1-u) \end{pmatrix}$$
(3.24)

that represents the right hand side of the system (3.21). The linearization about the equilibrium $(u, v) = (1, 0)$ is then given by

$$\frac{d\tilde{u}}{d\xi} = \tilde{v},$$
$$\frac{d\tilde{v}}{d\xi} = \tilde{u} - c\tilde{v},$$
(3.25)

since the Jacobian at this point is

$$J(1, 0) = \begin{pmatrix} 0 & 1 \\ 1 & -c \end{pmatrix}.$$
(3.26)

The eigenvalues of $J(1,0)$ are

$$\lambda_{1,2}(1,0) = -\frac{c}{2} \pm \frac{\sqrt{c^2 + 4}}{2}, \tag{3.27}$$

therefore $(1,0)$ is a saddle, since one of the eigenvalues is positive and the other is negative.

The Jacobian at the equilibrium $(u,v) = (0,0)$ is

$$J(0,0) = \begin{pmatrix} 0 & 1 \\ -1 & -c \end{pmatrix}. \tag{3.28}$$

Therefore the linearization of (3.19) at $(0,0)$ is

$$\begin{aligned} \frac{d\tilde{u}}{d\xi} &= \tilde{v}, \\ \frac{d\tilde{v}}{d\xi} &= -\tilde{u} - c\tilde{v}. \end{aligned} \tag{3.29}$$

The Jacobian $J(0,0)$ has the following eigenvalues

$$\lambda_{1,2}(0,0) = -\frac{c}{2} \pm \frac{\sqrt{c^2 - 4}}{2}. \tag{3.30}$$

Since $c > 0$, eigenvalues (3.30) have negative real part, therefore $(0,0)$ is a stable equilibrium. More specifically, equilibrium $(0,0)$ is a stable focus when $0 < c < 2$, it is a stable node when $c \geq 2$.

While we understand the characteristics of the flow within the local neighborhood of equilibria $(0,0)$ and $(1,0)$, which is an important ingredient in the analysis, it is not sufficient to conclude that there is a solution or an orbit that connects these two equilibria. The existence of the orbit that connects the equilibrium $(1,0)$ to equilibrium $(0,0)$ can be shown by using a trapping region argument. A trapping region is a closed connected set such that the vector field points inside the region at each point on the boundary of the region. Any trajectory that finds itself inside the region cannot leave it. To make a conclusion about the existence of a particular orbit one then a corollary the Poincaré-Bendixson Theorem [77, 93], which we informally formulate as follows: for a system of the first-order differential equations with a smooth vector field, if there is a closed region where the trajectories are confined, then the region contains either an equilibrium point or a closed orbit and the trajectories in the region either converge to an equilibrium point or wrap around the closed orbit.

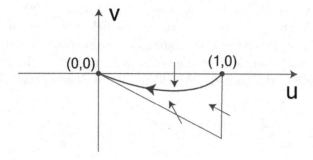

Figure 3.6 The trapping region and the trajectory trapped in it.

For our system, we consider a triangular region in the (u, v)-plane that consists of a horizontal segment connecting $(0, 0)$ with $(1, 0)$ along the u-axis, a vertical segment that goes from $(1, 0)$ along the line $u = 1$, and a segment given by $v = -\beta u$, where the constant β, $0 < \beta < 1$, is not defined yet. Since β is not defined, the third vertex $(1, -\beta)$ of the triangle is not defined yet either. The region is illustrated on Figure 3.6.

Let us look closely at the vector field generated by the system (3.21) along the sides of this simple triangular region. Along the side of the triangle with $v = 0$, since u has values between 0 and 1,

$$\frac{du}{d\xi} = 0,$$
$$\frac{dv}{d\xi} = -u(1 - u) < 0. \tag{3.31}$$

Therefore the vector field points vertically down.

Along the vertical side $u = 1$, since v has negative values, we have

$$\frac{du}{d\xi} = v < 0,$$
$$\frac{dv}{d\xi} = -cv > 0. \tag{3.32}$$

Therefore the vector field points up and to the left, which implies that the vector field points inside the region along this side of the triangle. We note that the position of the third vertex $(1, -\beta)$ does influence the direction of the vector field on this side.

Let us now consider the third side of the triangle along the line $v = -\beta u$. Clearly, $\frac{du}{d\xi} = v < 0$, so the vector field points to the left, but in this case more precise understanding of the slope of the vector field is needed. We want the slope of the vector field to be less than the slope $-\beta$ of the side of the triangle. We calculate the slope of the vector field using the chain rule as

$$\frac{dv}{du} = \frac{dv/d\xi}{du/d\xi} = \frac{-cv - u(1-u)}{v} = -c - \frac{u(1-u)}{v}. \tag{3.33}$$

Along the line $v = -\beta u$, this slope is

$$\frac{dv}{du} = -c - \frac{u(1-u)}{-\beta u} = -c + \frac{1}{\beta}(1-u). \tag{3.34}$$

Further, we notice that for any $0 < u \leq 1$,

$$\frac{dv}{du} = -c + \frac{1}{\beta}(1-u) < -c + \frac{1}{\beta}. \tag{3.35}$$

Let us find $\beta > 0$ such that $-c + \frac{1}{\beta} \leq -\beta$, which is equivalent (since $\beta > 0$) to

$$\beta^2 - c\beta + 1 \leq 0. \tag{3.36}$$

In that case, the vector field along $v = -\beta u$ is steeper than $v = -\beta u$ and, therefore, points inside of the region.

We notice that this equality does not hold when $c < 2$. Indeed, if $c < 2$, then the quadratic equation (3.36) has no real roots.

When $c = 2$, (3.36) has exactly one solution given by $\beta = 1$. Since the inequality (3.35) is strict everywhere on $0 < u \leq 1$, that implies that $\frac{dv}{du} < -\beta = -1$ everywhere along the line except for the vertex $(0,0)$.

When $c > 2$, it is easy to see that $\beta = 1$ still works. The quadratic inequality (3.36) holds for $\beta = 1$ or any β such that

$$\frac{c - \sqrt{c^2 - 4}}{2} \leq \beta \leq \frac{c + \sqrt{c^2 - 4}}{2}. \tag{3.37}$$

We summarize as follows: when $c \geq 2$, the vector field points inside of the triangular region as long as β that satisfies the condition (3.37), for example, when $\beta = 1$.

An alternative, geometric argument leading to the condition (3.37) may be made using the concept of the vector product. The parametric equation for the segment connecting $(0,0)$ and $(1, -\beta)$ is

$$\begin{pmatrix} u \\ v \end{pmatrix} = t \begin{pmatrix} -1 \\ \beta \end{pmatrix}, \quad 0 \leq t \leq 1. \tag{3.38}$$

The vector field at every point of this segment is then given by

$$\begin{pmatrix} -\beta t \\ c\beta t - t(1-t) \end{pmatrix}.$$ (3.39)

This vector field points inside of the triangle if and only if it points to the left of the vector that connects the vertex $(1, -\beta)$ with $(0,0)$. We add another dimension, perpendicular to the plane (u, v) and use the standard basis $(\mathbf{i}, \mathbf{j}, \mathbf{k})$ to reformulate this necessary and sufficient condition as follows: for any $0 < t < 1$ there is a positive constant α such that $\mathbf{n} \times (-1, \beta, 0)^T = \alpha \mathbf{k}$, where

$$\mathbf{n} = (-\beta t, c\beta t - t(1-t), 0)^T.$$ (3.40)

Let us calculate

$$\mathbf{n} \times (-1, \beta)^T = \det \begin{pmatrix} \mathbf{i} & \mathbf{j} & \mathbf{k} \\ -\beta t & c\beta t - t(1-t) & 0 \\ -1 & \beta & 0 \end{pmatrix} = t(-\beta^2 + c\beta - (1-t))\mathbf{k}.$$ (3.41)

Since $0 < t < 1$, we have that $\alpha = t(-\beta^2 + c\beta - (1-t)) > 0$ if and only if $\beta^2 - c\beta + (1-t) < 0$. Clearly, for $0 < t < 1$,

$$\beta^2 - c\beta + (1-t) < \beta^2 - c\beta + 1.$$ (3.42)

For $\beta > 0$ such that $\beta^2 - c\beta + 1 \leq 0$, then $\alpha = t(-\beta^2 + c\beta - (1-t)) < 0$, for any $0 < t < 1$. It is easy to see that $\beta^2 - c\beta + 1 \leq 0$ for any value of β that satisfies the condition (3.37). This completes the alternative argument.

We now may conclude that, for $c \geq 2$, the triangular region is a trapping region for any solutions that start inside it. Next, we concentrate at the vertex $(1,0)$, which we know is a saddle. We calculate the eigenvector that corresponds to the positive eigenvalue

$$\lambda_1(1,0) = -\frac{c}{2} + \frac{\sqrt{c^2 + 4}}{2},$$ (3.43)

to be

$$\begin{pmatrix} -1 \\ -\lambda_1 \end{pmatrix}.$$ (3.44)

Since $\lambda_1 > 0$, the eigenvector, when placed to start at the equilibrium $(1,0)$, points into the triangular region described above. The half-line

$$\begin{pmatrix} 0 \\ 1 \end{pmatrix} + t \begin{pmatrix} -1 \\ -\lambda_1 \end{pmatrix}, \quad t > 0,$$ (3.45)

is a trajectory to the linear system (3.25). In the original nonlinear system (3.23) then there is a trajectory for which the vector (3.44) is a tangent vector. Therefore, this trajectory, which has $(1,0)$ as a limit at $-\infty$, enters the trapping region.

Our calculation shows that this trajectory cannot leave the region through any of the sides of the triangle. There is no closed orbits in the region because the u-component is monotonically decreasing since $\frac{du}{d\xi} = v < 0$ inside of the region. Therefore, the trajectory has to converge as $\xi \to \infty$ to an equilibrium. That equilibrium cannot be $(1,0)$ because of the monotonic decrease of the u-component, so it must be the equilibrium $(0,0)$. Therefore there is an orbit that asymptotically connects $(1,0)$ to $(0,0)$ as ξ changes from $-\infty$ to ∞.

3.3.1 Existence of fronts in Nagumo equation

We consider the Nagumo equation (1.41) discussed earlier in Section 1.5.2,

$$u_t = u_{xx} + u(1-u)(u-a), \tag{3.46}$$

where we assume that the constant a satisfies $0 < a < \frac{1}{2}$. The equilibrium points of (3.46) are $u = 0$, $u = a$, and $u = 1$.

To capture traveling waves, we introduce in the equation (3.46) a moving frame $\xi = x - ct$,

$$u_t = u_{\xi\xi} + cu_\xi + u(1-u)(u-a), \tag{3.47}$$

and, after setting the time derivative equal to zero, obtain the traveling wave ordinary differential equation,

$$0 = u'' + cu' + u(1-u)(u-a), \tag{3.48}$$

where the prime denotes differentiation with respect to ξ. Let us focus on the front solutions that satisfy the following asymptotic conditions

$$\begin{aligned} \lim_{\xi \to -\infty} u(\xi) &= 0, \\ \lim_{\xi \to \infty} u(\xi) &= 1. \end{aligned} \tag{3.49}$$

We cast the ordinary differential equation (3.48) as a system of the first-order equations

$$\begin{aligned} \frac{du}{d\xi} &= v, \\ \frac{dv}{d\xi} &= -cv - u(1-u)(u-a). \end{aligned} \tag{3.50}$$

This system has the equilibrium points $(a, 0)$, $(0, 0)$, and $(1, 0)$. The latter two points are associated with the conditions (3.49). We start our analysis by linearizing the system (3.50) about the equilibrium points. To do so, we calculate the Jacobian for the vector-function

$$\begin{pmatrix} f_1(u, v) \\ f_2(u, v) \end{pmatrix} = \begin{pmatrix} v \\ -cv - u(1 - u)(u - a) \end{pmatrix}. \tag{3.51}$$

It is easy to see that the Jacobian is given by the matrix

$$J(u, v) = \begin{pmatrix} 0 & 1 \\ 3u^2 - 2(1 - a)u + a & -c \end{pmatrix}. \tag{3.52}$$

At the equilibrium $(u, v) = (0, 0)$ we have

$$J(0, 0) = \begin{pmatrix} 0 & 1 \\ a & -c \end{pmatrix}. \tag{3.53}$$

The linearization of the system (3.50) at this equilibrium is

$$\begin{aligned} \frac{d\tilde{u}}{d\xi} &= \tilde{v}, \\ \frac{d\tilde{v}}{d\xi} &= a\tilde{u} - c\tilde{v}. \end{aligned} \tag{3.54}$$

The Jacobian at $(0, 0)$ has the eigenvalues:

$$\lambda_{1,2}(1, 0) = \frac{-c \pm \sqrt{c^2 + 4a}}{2}. \tag{3.55}$$

Since $a > 0$, one of the eigenvalues is negative and one is positive. The equilibrium at $(0, 0)$ is a saddle. To satisfy the conditions (3.49) we have to look for the trajectories of (3.50) that approach $(0, 0)$ as $\xi \to -\infty$. These trajectories are tangent to the eigenvector corresponding to the positive eigenvalue

$$\lambda_1(0, 0) = \frac{-c + \sqrt{c^2 + 4a}}{2} \tag{3.56}$$

in the system (3.55), which we calculate to help ourselves visualize the phase portrait. The eigenvector is $(1, \lambda_1(0, 0))^T$.

At the equilibrium $(1, 0)$ we have

$$J(1, 0) = \begin{pmatrix} 0 & 1 \\ 1 - a & -c \end{pmatrix}. \tag{3.57}$$

The system (3.50) linearized at $(1,0)$ is

$$\frac{d\tilde{u}}{d\xi} = \tilde{v},$$
$$\frac{d\tilde{v}}{d\xi} = (1-a)\tilde{u} - c\tilde{v}. \tag{3.58}$$

The Jacobian at $(1,0)$ has the eigenvalues:

$$\lambda_{1,2}(1,0) = \frac{-c \pm \sqrt{c^2 + 4(1-a)}}{2}. \tag{3.59}$$

Since $a < 1$, one of the eigenvalues is negative and one is positive. The equilibrium at $(1,0)$ is also a saddle. At this equilibrium, to satisfy the conditions (3.49), we have to look for the trajectories of (3.50) that approach $(1,0)$ as $\xi \to +\infty$. These trajectories have as the tangent vector the eigenvector corresponding to the negative eigenvalue

$$\lambda_2(1,0) = \frac{-c - \sqrt{c^2 + 4(1-a)}}{2}. \tag{3.60}$$

This eigenvector is $(-1, -\lambda_2(1,0))^T$.

Observe that both equilibrium points are saddles regardless of the value and the sign of the speed parameter c. In particular, these are saddles when $c = 0$. Our proof of the existence consists of two steps. We first show that a heteroclinic orbit exists for the case $c = 0$. We then extend this result to the cases when c is sufficiently small, but nonzero.

The system (3.50) with $c = 0$

$$\frac{du}{d\xi} = v,$$
$$\frac{dv}{d\xi} = -u(1-u)(u-a) \tag{3.61}$$

is conservative [84]. Its total energy function is

$$H(u,v) = \frac{1}{2}v^2 + \int_0^u r(1-r)(r-a)dr, \tag{3.62}$$

where the first summand in H represents the kinetic energy and the second is the potential energy.

The total energy is constant along the trajectories of the system (3.61). Indeed, the derivative of H with respect to ξ is

$$\frac{d}{d\xi}H(u,v) = v\frac{dv}{d\xi} + u(1-u)(u-a)\frac{du}{d\xi}, \tag{3.63}$$

which, when substituting the expressions for the derivatives from (3.61), gives 0.

The vector field is symmetric with respect to the u axis. Indeed the substitution of $-v$ instead of v and $-\xi$ instead of ξ leaves the system invariant. In [84, Sec. 2.14, Theorem 3], this observation is listed along the observation about the nature of the equilibrium points of the system (3.61) and their classification as the critical points of the potential energy

$$P(u) = \int_0^u r(1-r)(r-a)dr \tag{3.64}$$

as a function of u. Notice that the critical points of $P(u)$ are simply zeros of the function $\frac{dP}{du} = u(1-u)(u-a)$. Theorem 3 from [84, Sec. 2.14] says that if $u = u_0$ is a local maximum of $P(u)$ then $(u_0, 0)$ is a saddle for the system (3.61). If it is a local minimum of $P(u)$ then it is a center for (3.61). If the point $(u_0, 0)$ is an inflection point with a horizontal tangent, then it is a cusp for (3.61). Perko [84] defines a cusp as a nonhyperbolic equilibrium characterized by the existence of only two separatrices which divide the complex plane in two sectors (a saddle, for example, has four separatrices and four sectors). The proof of Theorem 3 is left to the reader as an exercise in [84, Sec. 2.14, Ex. 9], and so we leave it as an exercise too.

Next, we take a closer look at (3.54). This system is a Newtonian system with potential energy (3.64). The dynamical properties of such systems are well known [84, p.173] or [41, p.48]. In particular, here the equilibria $A = (0,0)$, $C = (1,0)$ are saddles and $B = (a, 0)$ is a center. The phase portrait which is sketched in Figure 3.8 depends on the value of a.

The system (3.54) is reversible. Indeed, the system is invariant under the map $\{\xi \to -\xi, v \to -v\}$. The solution curves of (3.54) are given by $\frac{v^2}{2} - \frac{u^4}{4} + (1+a)\frac{u^3}{3} - a\frac{u^2}{2} = K$, or

$$v = \pm\sqrt{2\left(K + \frac{u^4}{4} - (1+a)\frac{u^3}{3} + a\frac{u^2}{2}\right)}.$$

It follows that when the two saddle points A and C (corresponding to the two maxima of $P(u)$) have the same energy $P(u^A) = P(u^C)$, then there exist heteroclinic orbits connecting them. Here by u^A, u^B, and u^C, we mean the u-components of the points (needed since P in (3.64) depend only on u), so $u^A = 0$, $u^B = a$, and $u^C = 1$. Note that $P(u^A) = 0$ and $P(u^C) = \frac{1}{12} - \frac{a}{6}$. Therefore, $P(u^A) = P(u^C)$ when $a = \frac{1}{2}$. Moreover, $P(u^C) > P(u^A) = 0$ when $a < \frac{1}{2}$, and $P(u^C) < 0$ when $a > \frac{1}{2}$.

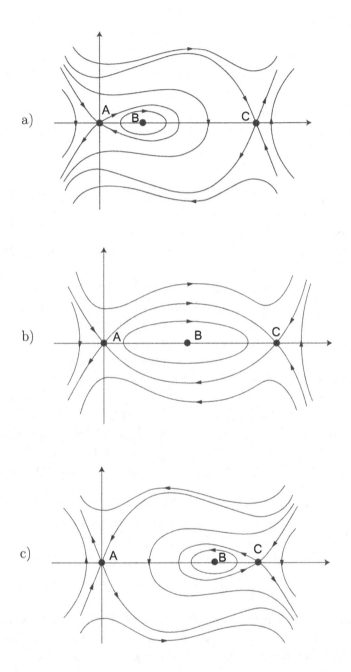

Figure 3.7 The phase portrait of (3.54) when $c = 0$: a) $P(u^C) > 0$, $0 < a < \frac{1}{2}$; b) $P(u^C) = 0$, $a = \frac{1}{2}$; c) $P(u^C) < 0$, $a > \frac{1}{2}$.

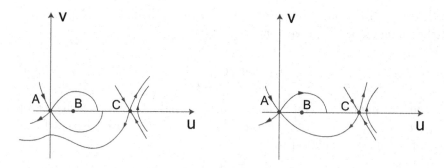

Figure 3.8 The phase portrait of (3.50) when $P(u^C) < 0$. Left: The homoclinic orbit from Figure breaks for $0 < c \ll 1$. Right: A heteroclinic orbit is formed for a value of $c = c^*(a)$.

Theorem 3.1 *Consider the dynamical system* (3.54) *with* $a > 0$. *Then*

- *for any* $0 < a < \frac{1}{2}$, *there exists a homoclinic orbit at the equilibrium* $A = (0,0)$;

- *for* $a = \frac{1}{2}$, *a unique up to reversibility symmetry, heteroclinic orbit that asymptotically connects* $A = (0,0)$ *and* $C = (1,0)$;

- *for any* $a > \frac{1}{2}$, *there exists a homoclinic orbit at the equilibrium* $C = (1,0)$.

Remark. In addition, for any $a > 0$, there exists a periodic orbit around the equilibrium $B = (a,0)$, but we concentrate on homoclinic and heteroclinic orbits asymptotic to A and C only.

3.3.2 Rotated vector fields and existence of a heteroclinic orbit between A and C for some $c \neq 0$

The vector field

$$(f_1(u,v), f_2(u,v)) = (v, -cv - u(1-u)(u-a))$$

that generates the flow of (3.50) defines a family of rotated vector fields on $\mathbb{R}^2 \setminus \{v = 0\}$ [83] with respect to the parameter c. It means that as the parameter c increases from 0 to ∞, the slope vectors at all points in (u,v)-plane, with the exception of points on the u-axis, rotate in unison in the same direction. This can be shown by establishing the dependance of the angle of the slope vector

$$\theta = \arctan\left(\frac{f_2}{f_1}\right) \tag{3.65}$$

on the parameter c. Note that θ is the angle between each vector $(f_1(u, v), f_2(u, v))$ and the u axis. The rate of change of θ as a function of c can be calculated to be

$$\frac{\partial \theta}{\partial c} = \frac{f_1 \frac{\partial f_2}{\partial c} - f_2 \frac{\partial f_1}{\partial c}}{f_1^2 + f_2^2} = -\frac{v^2}{f_1^2 + f_2^2}. \tag{3.66}$$

The derivative $\frac{\partial \theta}{\partial c}$ is strictly negative at each point of the plane, except at the points on the u-axis that is on $\mathbb{R}^2 \setminus \{v = 0\}$.

Since the equilibria of (3.50) do not depend on the parameter c and $\frac{\partial \theta}{\partial c}$ at all of the points of $\mathbb{R}^2 \setminus \{v = 0\}$, therefore due to [83, Theorem 5] about procession of separatrices in an analytic family of rotated vector fields with parameter c, upon increasing c from zero, the saddle separatrices of A and C rotate monotonically clockwise and meet forming a heteroclinic orbit, at a certain unique value of c. The heteroclinic orbit breaks upon increasing c further.

The system (3.50) is invariant under transformation $(u, v, t, c) \rightarrow (u, -v, -t, -c)$.

The following statement holds.

Theorem 3.2 *Consider the dynamical system* (3.50) *with* $a > 0$ *and* $c > 0$. *Then*

- *for any* $0 < a < \frac{1}{2}$, *there exists a unique* $c^*(a)$ *such that the stable manifold of A intersects the unstable manifold of C, thus forming a heteroclinic orbit asymptotically connecting A and C;*

- *for any* $a > \frac{1}{2}$, *there exists a unique* $c^*(a)$ *such that the unstable manifold of A intersects the stable manifold of C, thus forming a heteroclinic orbit.*

Remark. For $a = \frac{1}{2}$, heteroclinic connections formed by the stable manifold of A intersecting the unstable manifold of C and the stable manifold of C intersecting the unstable manifold of A exist for $c(\frac{1}{2}) = 0$ which are symmetric under transformation $(u, v, \xi, 0) \rightarrow (u, -v, -\xi, 0)$ as in Theorem 3.1. These heteroclinic orbits break when c is slightly varied.

3.4 EXISTENCE OF FRONTS IN SOLID FUEL COMBUSTION MODEL

In Section 2.2.2, we demonstrate that the existence of fronts in the solid fuel combustion model (2.40) is reduced to finding solutions to the traveling wave ordinary differential equation (2.45) with conditions (2.39). To prove the existence of such solutions we study the linearization of system (2.45) about the equilibria (or fixed points) $(1/\beta, 0)$ and $(0, 1)$.

The linearization of the system (2.45) about $(1/\beta, 0)$ is

$$\tilde{u}' = -c\tilde{u} - \frac{c}{\beta}\tilde{y},$$

$$\tilde{y}' = \frac{\beta}{c} e^{-\beta}\tilde{y}. \tag{3.67}$$

The Jacobian at $(1/\beta, 0)$

$$J(1/\beta, 0) = \begin{pmatrix} -c & -\frac{c}{\beta} \\ 0 & \frac{\beta}{c} e^{-\beta} \end{pmatrix} \tag{3.68}$$

has two nonzero eigenvalues of different signs: $-c$ and $\frac{\beta}{c} e^{-\beta}$, so it is a hyperbolic equilibrium, more precisely, a saddle. Any solution that satisfies (2.39) has to follow the unstable manifold of the equilibrium $(1/\beta, 0)$ as $\xi \to -\infty$. The tangent to the unstable manifold is spanned by the eigenvector corresponding to the positive eigenvalue $\frac{\beta}{c} e^{-\beta}$. That eigenvector is

$$\mathbf{v}_u \equiv (-c^2, \beta c^2 + \beta^2 e^{-\beta}) \tag{3.69}$$

or any of its nonzero scalings.

On the other hand, the linearization of the system (2.40) about $(0, 1)$ is

$$\tilde{u}' = -c\tilde{u} - \frac{c}{\beta}\tilde{y},$$

$$\tilde{y}' = 0. \tag{3.70}$$

The Jacobian at $(0, 1)$

$$J(0, 1) = \begin{pmatrix} -c & -\frac{c}{\beta} \\ 0 & 0 \end{pmatrix} \tag{3.71}$$

has one negative eigenvalue $-c$ and one 0 eigenvalue, so it is a non-hyperbolic equilibrium. The tangent to the stable manifold at $(0, 1)$ is given by the vector $(1, 0)$.

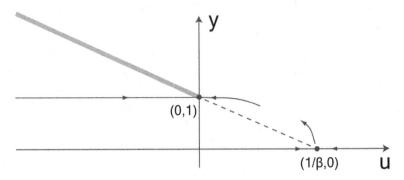

Figure 3.9 Phase portrait of (2.45). There is a unique value of c for which the unstable manifold of the equilibrium $(1/\beta, 0)$ and the stable manifold of the equilibrium $(0, 1)$ meet.

It is known from [97] that there is a unique value of $c > 0$ for which the system (2.40) has a solution that asymptotically connects the equilibria $(1/\beta, 0)$ and $(0, 1)$. This observation follows from the geometrical observation illustrated in Figure 3.9. In the limit $c = 0$, the tangent to the unstable manifold $(1/\beta, 0)$ is vertical. For very large c, it is aligned with $(1, -\beta)$, which can be seen by dividing $(c^2, -\beta c^2 - \beta^2 e^{-\beta})$ by c^2 and taking a limit as $c \to \infty$. The vector $(1, -\beta)$ is co-linear with the central manifold of $(0, 1)$. The angle between the unstable manifold and the positive opening of the y-axis is given by formula (3.65). Its derivative with respect to c is

$$\frac{\partial \theta}{\partial c} = \frac{\frac{c}{\beta}(1 - y - \beta u)\left(-\frac{\beta}{c^2} y\, e^{-1/u}\right) - \frac{\beta}{c} y\, e^{-1/u} \frac{1}{\beta}(1 - y - \beta u)}{\left(\frac{c}{\beta}(1 - y - \beta u)\right)^2 + \left(\frac{\beta}{c} y\, e^{-1/u}\right)^2}. \tag{3.72}$$

The sign of $\frac{\partial \theta}{\partial c}$ is determined by the sign of its numerator. The numerator can be simplified to

$$-\frac{2}{c}(1 - y - \beta u)\, y\, e^{-1/u}. \tag{3.73}$$

The region above the central manifold is characterized by the inequality $y > 1 - \beta u$, hence the expression (3.73) is strictly positive. Therefore θ is increasing monotonically in that region, and so the unstable manifold moves monotonically, counterclockwise. There will be exactly one value of c, say $c = c^*$, such that the unstable manifold of manifold of the fixed point $(1/\beta, 0)$ meets the stable manifold of the fixed point $(0, 1)$,

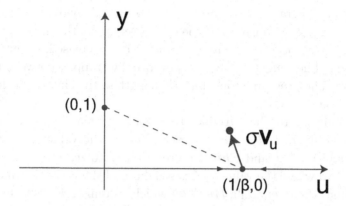

Figure 3.10 Illustration of the initial condition used in (3.74) to perform the shooting from the fixed point $(1/\beta, 0)$ in the unstable direction.

thus, forming a heteroclinic orbit that asymptotically connects these equilibria.

We now obtain this heteroclinic orbit numerically. The numerics relies on the analysis performed to prove existence. We use what is called a "shooting method" (see for example [85]). The solution and the corresponding value $c = c^*$ are obtained in the (u, y)-plane by "shooting" from the fixed point $(1/\beta, 0)$ in the unstable direction given in (3.69) for various values of c until a connection to the fixed point $(0, 1)$ is found. For each value of c, the shooting is done by integrating system (2.45) for initial conditions that are close to the fixed point $(1/\beta, 0)$ in the direction of $\mathbf{v}_u$ given in (3.69). More precisely, the initial condition is given to be

$$(u(0), y(0)) = (1/\beta, 0) + \sigma \mathbf{v}_u, \tag{3.74}$$

where σ is a small constant (see Figure 3.10).

The MATLAB scripts doing those computations are located on Github at the address specified in [4]. The scripts are in the subfolder HighLewis/WaveComputation. In this folder, the file integrated_ode.m contains the expressions defining system (2.45), the file integrated_solve.m numerically solves system (2.45) given the values of c and β and initial condition *initial*. The file integrated_find_c.m only requires the value of β to be specified and it uses the script integrated_find_c.m to numerically solve system (2.45) for various values of c. The general idea is based on what is illustrated in Figure 3.9.

The script integrates the system (2.45) for initial conditions of the form (3.74) for a given value of c. If the orbit overshoots the fixed point (that is, $y > 1$ when $u = 0$), then the value of c is increased, otherwise it is decreased. The sought value of c is eventually trapped between two real numbers. The program stops until the length of that interval is less than the variable *precision*.

In addition to the variable *precision*, the three scripts described above contain *initial_scale*, corresponding to the value of σ in (3.74), the variables *c_low* and *c_high*, corresponding to the initial endpoint of the interval where the value of c^* is sought, and the variable *final*, which determines the interval $[0, final]$ on which system (2.45) is to be numerically solved. The variable *initial_scale* is taken to be small. The best choice for the value of *initial_scale* is determined by numerical experimentations. For example, a value taken to be too small might require the end point of the ξ interval to be larger and cause the program to run longer to obtain the solution. A value of *initial_scale* taken to be too large might prevent the program from computing the front solution. The requirement on the variables *c_low* and *c_high* is that the velocity c^* be in the interval $[c\_low, c\_high]$. If it is not, the program will end up with the final value of c be equal to one of *c_low* or *c_high*, and the graph of the obtained solution will not have the desired appearance. In such a case, a larger interval is required. The value of the variable *final* does not matter as long as it is positive. The script integrated_solve.m will increase that value if the obtained solution is well-defined and the first component u of the solution is not closed enough to zero. This is done in the following lines of the script.

```
final = final * 2;
        disp(['extending to ', num2str(final)]);
        sol = odextend(sol, [], final);
```

Furthermore, there is a limit that the variable *final* can take set by the following line.

```
if final > 260
        disp('giving up');
        break;
```

This upper limit of 260 may need to be changed due to different choices in the other parameters. Note that for system (2.45), there is the additional difficulty that the second equation is not well-defined when $u = 0$ due to the expression $e^{-1/u}$. The program integrated_solve.m thus contains the

following lines that make sure the obtained solution does not include the
MatLab values NaN or Inf, due to the fact that u gets too close to zero.

```
if isnan(sol.y(2,end)) | isinf(sol.y(2,:))
       test=1;
       indf=find(isnan(sol.y(2,:)) | isinf(sol.y
          (2,:)), 1,'first')-1;
       final=sol.x(indf);
       sol = ode45(ode, [0,final], initial,
          options);
       disp('allo');
   elseif sol.y(1,end)<10^(-1)
       test=1;
```

The three scripts integrated_ode.m, integrated_solve.m, and integrated_find_c.m described above find the speed c^*, compute the solution, and graph it. For example, the solution is obtained for $\beta = 1$ by the following command line:

```
[c, front, sol] = integrated_find_c(1)
```

The velocity is found to be $c = 0.5707$. Figure 3.11 illustrates the obtained front. Figure 3.12 illustrates the solutions in the (u, y)-plane calculated for the three values velocity parameter: $c = 0.65$, $c = 0.5$, and the front velocity $c = c^* = 0.5707$. That last figure shows how the resulted solution overshoots when $c > c^*$, while it undershoots when $c < c^*$.

3.5 WAVETRAINS

In this section we will consider the wavetrain solutions of the Brusselator model

$$p_\tau = \epsilon_p p_{zz} + k_1 - (k_2 + k_4)p + k_3 p^2 q,$$
$$q_\tau = \epsilon_q q_{zz} + k_2 p - k_3 p^2 q. \tag{3.75}$$

The Brusselator model was derived in [86] as a theoretical model for the Belousov-Zhabotinsky reaction which is a chemical reaction that supports sustained oscillations. In this model, parameters ϵ_p, $\epsilon_q > 0$ are the diffusion constants and k_1, k_2, k_3, and k_4 are positive constants that come from the equations of the chemical kinetics.

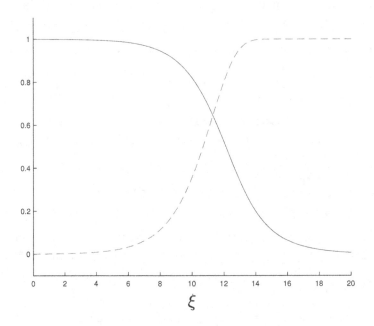

Figure 3.11 Front solution of system (2.40) in the case $\beta = 1$ with obtained velocity $c = 0.5707$. The solid line corresponds to u and the dashed line to y.

Let us transform the system (3.75) using the following scalings

$$
u = \frac{k_1}{k_4}p, \quad t = k_4\tau, \quad x = \sqrt{k_4}z,
$$
$$
\alpha = \frac{k_3 k_1}{k_4^2}, \quad \beta = \frac{k_2}{k_4}, \quad \eta = \frac{k_1}{k_4}.
$$

(3.76)

The new parameters α, β and η are positive. The goal of this transformation is to rewrite (3.75) in a non-dimensional form

$$
u_t = \epsilon_p u_{xx} + 1 - (\beta + 1)u + \alpha u^2 v,
$$
$$
v_t = \epsilon_q v_{xx} + \eta(-\alpha u^2 v + \beta u).
$$

(3.77)

We further set $\epsilon_p = \epsilon$, and $\epsilon_q = r\epsilon$, where we think of r as a fixed constant representing the ratio of ϵ_q and ϵ_p,

$$
u_t = \epsilon u_{xx} + 1 - (\beta + 1)u + \alpha u^2 v,
$$
$$
v_t = r\epsilon v_{xx} + \eta(-\alpha u^2 v + \beta u).
$$

(3.78)

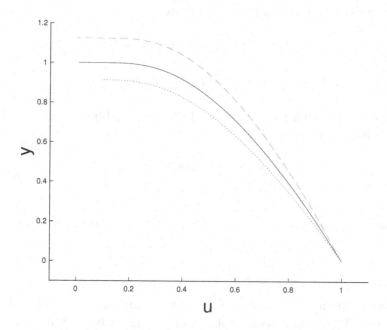

Figure 3.12 Phase plane illustration of solutions of system (2.45) in the case $\beta = 1$ in three cases: $c = 0.65$ (dotted line), $c = 0.5$ (dashed line) and the front obtained velocity $c = 0.5707$ (solid line).

In what follows we focus on the situations when ϵ as positive and small, that is, $0 < \epsilon \ll 1$.

To study wavetrain solutions we cast (3.78) using traveling wave coordinates $\xi = x - ct$ as

$$u_t = \epsilon u_{\xi\xi} + c u_\xi + 1 - (\beta + 1)u + \alpha u^2 v,$$
$$v_t = r\epsilon v_{\xi\xi} + c v_\xi + \eta(-\alpha u^2 v + \beta u), \qquad (3.79)$$

and seek the wavetrains as the stationary periodic solutions. Therefore, the wavetrains are solutions of the nonlinearly coupled system of ordinary differential equations

$$0 = \epsilon u'' + c u' + (1 - (\beta + 1)u + \alpha u^2 v),$$
$$0 = r\epsilon v'' + c v' + \eta(-\alpha u^2 v + \beta u). \qquad (3.80)$$

Note that when $\epsilon = 0$, the system above reduces to the following

$$0 = c u' + (1 - (\beta + 1)u + \alpha u^2 v),$$
$$0 = c v' + \eta(-\alpha u^2 v + \beta u), \qquad (3.81)$$

which can be written as a first-order system

$$u' = -\frac{1}{c}(1 - (\beta + 1)u + \alpha u^2 v),$$
$$v' = -\frac{\eta}{c}(-\alpha u^2 v + \beta u). \tag{3.82}$$

Equilibria of the dynamical system (3.82) are found by solving the algebraic system of equations

$$0 = 1 - (\beta + 1)u + \alpha u^2 v,$$
$$0 = -\alpha u^2 v + \beta u. \tag{3.83}$$

The unique equilibrium $(1, \frac{\beta}{\alpha})$ of (3.83) is the intersection of two curves

$$v = \frac{(\beta + 1)u - 1}{\alpha u^2} \quad \text{and} \quad v = \frac{\beta}{\alpha u}. \tag{3.84}$$

The equilibrium $(1, \frac{\beta}{\alpha})$ can be found by setting the right sides of the equations (3.84) equal to each other and solving for u. After finding that $u = 1$, we plug $u = 1$ into the second equation and find that $v = \frac{\beta}{\alpha}$.

We then study the linear approximation of (3.82) near this equilibrium. Linearization of (3.82) is given by the Jacobian of the vector-function

$$f(u, v) = \frac{1}{c} \begin{pmatrix} -1 + (\beta + 1)u - \alpha u^2 v \\ \eta(\alpha u^2 v - \beta u) \end{pmatrix}, \tag{3.85}$$

which is

$$J(u, v) = \frac{1}{c} \begin{pmatrix} \beta + 1 - 2\alpha uv & -\alpha u^2 \\ \eta(\beta - 2\alpha uv) & \eta \alpha u^2 \end{pmatrix}. \tag{3.86}$$

The Jacobian evaluated at $(u, v) = (1, \frac{\beta}{\alpha})$ is

$$J\left(1, \frac{\beta}{\alpha}\right) = \frac{1}{c} \begin{pmatrix} 1 - \beta & -\alpha \\ \eta\beta & \eta\alpha \end{pmatrix}. \tag{3.87}$$

The eigenvalues of the Jacobian are

$$\lambda_{1,2} = \frac{\eta\alpha + 1 - \beta \pm \sqrt{(\eta\alpha + 1 - \beta)^2 - 4\alpha\eta}}{2c}. \tag{3.88}$$

We treat $\eta > 0$ as the bifurcation parameter and seek a transition between qualitatively different situations as η varies. Values of η when such transitions occur are called bifurcation values or bifurcation points. One such transition occurs at

$$\eta_H = \frac{\beta - 1}{\alpha}. \tag{3.89}$$

Indeed, when $\eta = \eta_H$ the eigenvalues (3.88) are purely imaginary. When η is sufficiently close to η_H, then the eigenvalues are complex. The real parts of the eigenvalues change their sign as η crosses η_H. The change in the sign of the real parts of the eigenvalues results in the change of the stability of the equilibrium. The equilibrium changes from stable to unstable focus. At the bifurcation point the nature of the equilibrium is not determined by the linear approximation of the system, instead the interplay between the linear and nonlinear terms plays a decisive role. The appearance of a periodic orbit through a change in the stability of an equilibrium point is called a Hopf bifurcation.

We know that by definition η, β and α are positive parameters. To guarantee that the value of η_H is positive, we additionally assume that

$$\beta > 1. \tag{3.90}$$

More analysis may be done to obtain additional information about the bifurcation of the periodic orbits and their stability. A periodic orbit is called asymptotically stable if all nearby solutions converge to it as the independent variable goes to $+\infty$. It is called stable in nearby orbits stay nearby. If it is not stable or asymptotically stable, then it is called unstable. For a class of systems with polynomial nonlinearities, the Hopf Bifurcation Theorem [84] uses a quantity called Liapunov number to identify when periodic orbits bifurcate and their stability. More precisely, let us consider systems of the form

$$\begin{aligned} u_1' &= a u_1 + b v_1 + p(u_1, v_1), \\ v_1' &= c u_1 + d v_1 + q(u_1, v_1), \end{aligned} \tag{3.91}$$

where $ad - bc > 0$ and

$$p(u_1, v_1) = \sum_{j+j \geq 2} a_{ij} u_1^i v_1^j, \quad q(u_1, v_1) = \sum_{j+j \geq 2} b_{ij} u_1^i v_1^j.$$

We note that the assumptions on the system guarantee that the origin $(u_1, v_1) = (0, 0)$ is an isolated equilibrium in the (u_1, v_1)-plane.

The Liapunov number for (3.91) is defined by the formula

$$
\sigma = \frac{-3\pi}{2b\Delta^{3/2}} \Big\{ [ac(a_{11}^2 + a_{11}b_{02} + a_{02}b_{11}) + ab(b_{11}^2 + a_{20}b_{11} + a_{11}b_{02})
$$
$$
+ c^2(a_{11}a_{02} + 2a_{02}b_{02}) - 2ac(b_{02}^2 - a_{20}a_{02}) - 2ab(a_{20}^2 - b_{20}b_{02})
$$
$$
- b^2(2a_{20}b_{20} + b_{11}b_{20}) + (bc - 2a^2)(b_{11}b_{02} - a_{11}a_{20})]
$$
$$
- (a^2 + bc)[3(cb_{03} - ba_{30}) + 2a(a_{21} + b_{12}) + (ca_{12} - bb_{21})] \Big\}.
$$

$$(3.92)$$

The Hopf Bifurcation Theorem [84] relates the value of the parameter $\mu = a+d$ with the existence of periodic orbits. When $\mu = 0$, the Jacobian of the system has purely imaginary eigenvalues. Indeed, the eigenvalues are roots of the polynomial $(a - \lambda)(d - \lambda) - bc = 0$, which when $a + d = 0$ is reduced to $\lambda^2 + ad - bc = 0$, and so $\lambda = \pm i\sqrt{ad - bc}$ since $ad - bc > 0$.

The Hopf Bifurcation Theorem states that if $\sigma \neq 0$, then a Hopf bifurcation occurs at the origin at $\mu = 0$ which is then called the bifurcation value. Moreover, if $\sigma < 0$, then a unique stable periodic orbit appears when μ increases from 0, and if $\sigma > 0$, then a unique unstable periodic orbit appears when μ decreases from 0. We now apply this theorem to the system (3.82).

In order to cast the system (3.82) in the form (3.91), we shift the equilibrium $(u, v) = (1, \frac{\beta}{\alpha})$ to the origin by means of the linear coordinate transformation

$$
u = u_1 + 1, \quad v = v_1 + \frac{\beta}{\alpha}, \tag{3.93}
$$

and obtain an equivalent to (3.82) system

$$
\dot{u}_1 = \frac{(1 - \beta)}{c}u_1 - \frac{\alpha}{c}v_1 - 2\frac{\alpha}{c}u_1v_1 - \frac{\alpha}{c}u_1^2v_1,
$$
$$
\dot{v}_1 = \frac{\eta\beta}{c}u_1 + \frac{\eta\alpha}{c}v_1 + \frac{\eta\beta}{c}u_1^2 + +2\frac{\eta\alpha}{c}u_1v_1 + \frac{\eta\alpha}{c}u_1^2v_1.
$$

$$(3.94)$$

For this system, the bifurcation parameter μ that is used in the Hopf Bifurcation Theorem is

$$
\mu = \frac{1 - \beta}{c} + \frac{\eta\alpha}{c}. \tag{3.95}
$$

It is easy to see that $\mu = 0$ when $\eta = \eta_H$, $\mu > 0$ when $\eta > \eta_H$, and $\mu < 0$ when $\eta < \eta_H$, where η_H is defined by (3.89).

Let us calculate the Liapunov number for the system (3.94). Comparing the system (3.94) with (3.91), we see that

$$a = \frac{1 - \beta}{c}, \qquad b = -\frac{\alpha}{c}, \qquad c = \frac{\eta}{c}, \qquad d = \frac{\eta\alpha}{c}, \tag{3.96}$$

and the coefficients of polynomial functions p and q are

$$
\begin{aligned}
&a_{11} = -2\frac{\alpha}{c}, \quad a_{20} = -\frac{\beta}{c}, \quad a_{21} = -\frac{\alpha}{c}, \\
&a_{02} = 0, \quad a_{12} = 0, \quad a_{30} = 0, \quad a_{03} = 0, \\
&b_{11} = -2\frac{\eta\alpha}{c}, \quad b_{20} = -\frac{\eta\beta}{c}, \quad b_{21} = -\frac{\eta\alpha}{c}, \\
&b_{02} = 0, \quad b_{12} = 0, \quad b_{30} = 0, \quad b_{03} = 0.
\end{aligned}
\tag{3.97}
$$

Therefore, formula (3.92) gives

$$\sigma = \frac{3\pi(2\beta - 1)(\beta + 1)}{2\sqrt{\beta - 1}}, \tag{3.98}$$

therefore the condition (3.90) implies that $\sigma > 0$. It follows from the Hopf Bifurcation Theorem that a unique unstable limit cycle will bifurcate from the fixed point when $\eta < \eta_H = \frac{\beta - 1}{\alpha}$.

Recall that the system (3.82) is obtained from the system (3.80) by setting $\epsilon = 0$. In some sense, (3.82) is a limit of the system (3.80) as $\epsilon \to 0$, but this limit is singular as it changes the dimensions of the underlying phase space. When written as a system of equations of the first order, the system (3.80) is four-dimensional, while (3.82) is only two-dimensional. Because of this reduction in the dimension, it is not immediately clear whether or not there are orbits in (3.80) which converge as $\epsilon \to 0$ to the orbits in (3.82) existence of which we have shown. Additional analytical work is necessary to show their existence, which is outside of the scope of this book, but in general, if one could show existence of an orbit in a limit of a system, there is a good chance that there is an orbit which is an ϵ-order perturbation of the orbit in the limit $\epsilon \to 0$ since periodic orbits are structurally stable orbits, therefore persist under small perturbations. Persistence of periodic orbits for small ϵ in the Brusselator system can be shown by following the approach taken in [38] or in [20].

Note that in Chapter 4, dealing with the topic of stability of traveling waves, we restrict ourselves to pulses and fronts. This is because the methods and techniques needed for analysis of stability properties of wavetrain solutions require graduate level preparation, therefore the question of stability of such solutions is outside the scope of this book.

CHAPTER 4

Stability of fronts and pulses

4.1   STABILITY: INTRODUCTION

Stability is a fundamental concept in physics. An illustration of this concept is given by trying to make a pencil stand on its lead. In theory, it is possible but, in practice, because it is such an unstable state, it cannot be done. Any little perturbation would make the pencil fall on its side. In the example just described, the study is very simple and there is no need for a mathematical analysis to prove or disprove the stability of the system. The concept of stability carries over to solutions of differential equations such as the ones discussed in this book. In this context, the concept of stability takes an abstract form and its study involves some sophisticated mathematical tools. However, since solutions such as traveling waves can be representations of physical phenomena in applications, stability is an important concept. It describes the resilience of the solutions under perturbations. If a solution is stable, it is expected to keep its shape for a long period of time and be observed experimentally. In the case where there is instability, while the solution may not be persistent, a stability study is still important as it may shed light on structures that emerge under a perturbation.

In the case of traveling wave solutions, we have translational invariance, that is, if $u = u_0(x - ct)$ is a solution then so is the translate $u = u_0(x - ct + \xi_0)$, where ξ_0 is an arbitrary constant. Thus, when perturbed, a stable traveling wave solution is expected to stay near the family of all translations of the wave. In that case, the stability is called *orbital*. If, in addition to staying close, it settles at a translation of itself,

DOI: 10.1201/9781003147619-4

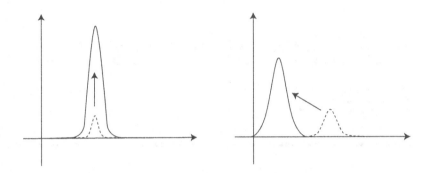

Figure 4.1 The left panel shows a schematic representation of the absolute instability. The panel on the right schematically illustrates the case where the perturbation is convected in one preferred direction.

it is called *nonlinear stability with an asymptotic phase*. If the wave is irrecoverably distorted by a perturbation, then it is called unstable.

The instability mechanisms on unbounded domains (i.e. for solutions defined on all of the real line) involve not only growth of perturbations but their propagation as well. If an initial perturbation grows in time but is simultaneously transported to infinity, and eventually dies out at each point in the space, then the instability is called convective. This is in contrast to an absolute instability which is manifested as growth of an initial perturbation at each spatial location in a coordinate frame that moves together with the wave (see Figure 4.1). Convective instability is an interesting type of instability because in point-wise sense (i.e. for each value of ξ) it is a "stable" regime. The concepts of absolute and convective instabilities were first introduced by R. J. Briggs in the study of the instabilities that occur when an electron beam is injected into a plasma [13]. The importance of these concepts spreads across different areas of applied science.

The stability analysis is a multistep process.

The first step consists in linearizing the equation about the traveling wave solution. As we will see, the linearization provides a linear equation that approximatively describes the behavior of a perturbation to the solution. The stability analysis is then based on the information about the location of the spectrum of a differential operator obtained from the linearization of the equation. The spectrum of a differential operator is an idea that generalizes the concept of eigenvalue for a matrix. In general, the spectrum of a differential operator consists of discrete eigenvalues of finite multiplicity and continuous spectrum. The stability analysis of traveling waves starts by identifying whether there exists an unstable spectrum . More precisely, a traveling wave is spectrally stable if the spectrum of the linearization is contained strictly in the open left half-plane of the complex plane with the exception of a simple eigenvalue at 0. Hence, what we call an unstable spectrum , is any element of the spectrum on the closed right side of the complex plane (that is, with the nonnegative real part) except the zero value.

The presence of an unstable spectrum implies the instability of the wave. The type of the unstable spectrum is also important. If there are unstable discrete eigenvalues of finite multiplicity then one may expect perturbations to the wave to grow exponentially in time. The solution will lose its coherent form. If it is the essential spectrum that is unstable, then additional analysis such as using exponential weights, finding the location of absolute spectrum (the spectrum that cannot be moved by exponential weights), or identifying which of the rest states is responsible for the instability may show whether the instability is convective or absolutely unstable [88].

With respect to the evolution of the perturbations to the wave in the original nonlinear system, a traveling wave is called nonlinearly stable with asymptotic phase if a solution that starts near the wave converges to one of the translations of the wave as time approaches ∞. Mathematical definition of the convective instability used exponentially weighted norms. The instability is called convective if there exists an exponentially weighted space (i.e. a function space in which the perturbations are multiplied by an exponential function) in which perturbations will converge to zero [87]. In other words, the perturbation may be growing in time in some way by itself, but multiplying it by some exponential function may curb this growth or even cause it to decay (see Figure 4.2).

The absence of the unstable spectrum does not necessarily guarantee the stability of the wave. So, there comes the second step in the stability analysis. Its purpose is to relate the spectral information to the nonlinear stability of the wave. If the linearized operator belongs to a special class of operators called sectorial [44], then a spectrally stable wave is also nonlinearly stable. But, generally speaking, it is not always true, or not always easy to see that the stability properties of the wave in the full nonlinear equation are dominated by the spectral properties of the linearized operator.

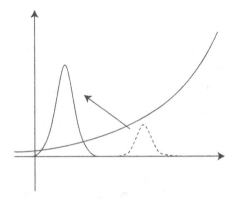

Figure 4.2 Assume that the perturbation increases but is at the same time transported to the left, to the $-\infty$. If we multiply the function that describes the perturbation by an exponential function with a carefully chosen positive rate, so it approaches zero at $-\infty$, then the growth of the perturbation may be balanced by the decay of the weight.

Two main situations when difficulties in the analysis arise are *when the linearized operator is not sectorial* and *when the wave has an essential spectrum extending to or across the imaginary axis*. The first situation is true for partially parabolic systems. In the second situation, the spectral information is not definitive. Traditionally, exponential weights have been used to achieve spectral stability, but there is no general theory that would provide the transition from the spectral to the nonlinear stability of the waves in the weighted spaces. Most of the nonlinearities do not behave well in weighted norms. Refined analysis is needed to study issues related to convection of small perturbation by the full nonlinear system.

Sections 4.2 and 4.3 address the spectral stability of fronts and pulses. Section 4.4 addresses their nonlinear stability in the case when the linearization about the wave has no unstable spectrum.

4.2   A HEURISTIC PRESENTATION OF SPECTRAL STABILITY FOR FRONT AND PULSE TRAVELING WAVE SOLUTIONS

In this section and the two following ones, we deal with the stability analysis of fronts and pulses (see Figure 1.1). As mentioned before, traveling waves that asymptotically at $+\infty$ and $-\infty$ connect two distinct spatially homogeneous equilibria are called fronts. If equilibria are the same, the wave is called a pulse. In the context of traveling waves, the equilibria are often called the rest states of the wave.

4.2.1 Eigenvalue problem

The intuitive idea for studying the stability of a solution to a partial differential equation is to add a "small" perturbation to a solution and verify if the perturbed solution stays close to the original solution or drifts away from it. The wording "linear stability" refers to stability study where one considers the contribution of the perturbation of the first degree only. In effect, this implies that the perturbation satisfies a linear equation. While this simplifies the study tremendously (a linear equation is generally simpler than a nonlinear one), it does have the disadvantage of being an approximation. However, it is often the starting point for a more thorough study.

We first describe the general notation. We consider a partial differential equation of the form

$$u_t = \mathcal{P}(\partial_x)u + \mathcal{N}(u), \tag{4.1}$$

where $\mathcal{P}$ is a polynomial function of the operator ∂_x. For example, it could be the following third-order operator

$$\mathcal{P}(\partial_x) = \partial_x^3 + \partial_x^2 + \partial_x + 1.$$

Other examples are given in (4.7) and (4.21) below. In (4.1), $\mathcal{N}$ is assumed to be an analytic function of u and its derivatives so, in principle, we should write $\mathcal{N}(u, u_x, u_{xx}, u_{xxx}, ...)$ in (4.1). However, we only write $\mathcal{N}$ as a function of u to simplify the notation. To generalize the notation (4.1) to the case of a system of equations, we can think of u and $\mathcal{N}$ as vector valued functions and $\mathcal{P}$ as a matrix function of ∂_x. A specific example in that case is given in (4.24) below. If the notation is a bit nebulous at this point, we will go over several examples and explain how each of those examples is of the form (4.1).

Since we are studying the stability of traveling wave solutions, we write (4.1) in the traveling wave variables (t, ξ), $\xi = x - ct$ as follows

$$u_t = cu_\xi + \mathcal{P}(\partial_\xi)u + \mathcal{N}(u). \tag{4.2}$$

Let $u(t, x) = u_0(\xi)$ be a traveling wave solution to the equation above. Since u_0 only depends on ξ, it satisfies the equation

$$cu_{0\xi} + \mathcal{P}(\partial_\xi)u_0 + \mathcal{N}(u_0) = 0. \tag{4.3}$$

We substitute the expression $u = u_0(\xi) + v(t, \xi)$ into (4.1), where v is thought of as a "small" perturbation to the solution u_0. Using the fact that $u = u_0(\xi)$ solves (4.1), we find

$$
\begin{aligned}
(u_0 + v)_t &= c(u_0 + v)_\xi + \mathcal{P}(\partial_\xi)\,(u_0 + v) + \mathcal{N}(u_0 + v), \\
v_t &= \underbrace{-u_{0t} + cu_{0\xi} + \mathcal{P}(\partial_\xi)u_0 + \mathcal{N}(u_0)}_{=0} \\
&\quad + cv_\xi + \mathcal{P}(\partial_\xi)v + \mathcal{N}(u_0 + v) - \mathcal{N}(u_0), \\
&= cv_\xi + \mathcal{P}(\partial_\xi)v + \mathcal{N}(u_0 + v) - \mathcal{N}(u_0),
\end{aligned}
\tag{4.4}
$$

where we were able to set the right side of the second line to zero because $u_0(\xi)$ does not depend on t and because it satisfies equation (4.3). We deal with the last two terms in (4.4) by using a Taylor expansion, since $\mathcal{N}$ is assumed to be analytic:

$$
\begin{aligned}
\mathcal{N}(u_0 + v) - \mathcal{N}(u_0) &= v \left(\frac{\partial \mathcal{N}}{\partial u} \right)\bigg|_{u=u_0} + v_\xi \left(\frac{\partial \mathcal{N}}{\partial u_\xi} \right)\bigg|_{u=u_0} \\
&\quad + v_{\xi\xi} \left(\frac{\partial \mathcal{N}}{\partial u_{\xi\xi}} \right)\bigg|_{u=u_0} + v_{\xi\xi\xi} \left(\frac{\partial \mathcal{N}}{\partial u_{\xi\xi\xi}} \right)\bigg|_{u=u_0} + \dots,
\end{aligned}
\tag{4.5}
$$

where the dots represent the terms that are of degree two or more in v and its derivatives. To simplify the notation, we only considered dependence of $\mathcal{N}$ on u and its first three derivatives. The linearization to the partial differential equation (4.2) then is obtained by substituting (4.5) into (4.4) and only keeping the terms that are linear in v and its derivatives to obtain

$$
\begin{aligned}
v_t &= cv_\xi + \mathcal{P}(\partial_\xi)v + v \left(\frac{\partial \mathcal{N}}{\partial u} \right)\bigg|_{u=u_0} + v_\xi \left(\frac{\partial \mathcal{N}}{\partial u_\xi} \right)\bigg|_{u=u_0} \\
&\quad + v_{\xi\xi} \left(\frac{\partial \mathcal{N}}{\partial u_{\xi\xi}} \right)\bigg|_{u=u_0} + v_{\xi\xi\xi} \left(\frac{\partial \mathcal{N}}{\partial u_{\xi\xi\xi}} \right)\bigg|_{u=u_0}.
\end{aligned}
\tag{4.6}
$$

Intuitively, the linearization helps us understand the evolution of a small perturbation to the solution $u = u_0(\xi)$. We say that the evolution is approximate because, when going from (4.4) to (4.6), terms were dropped at the price of losing exactness but with the reward of obtaining a linear equation. Hence, intuitively, if a solution to (4.6) starts (at $t = 0$) "small" but then grows in time, we expect the traveling wave solution to be unstable. But when dealing with the linear approximation (4.6), we only use the term linear instability. One might think it is easy to define what we mean by "small" when thinking of perturbations. However, perturbations are functions defined on the whole line $\mathbb{R}$. We will see later a concept that enables us to say how such an object can be considered small. Despite that, we now introduce the following definition in which we loosely use the term "small".

Definition 4.1 *We say that the traveling wave solution* $u = u_0(\xi)$ *to* (4.2) *is* linearly *stable, if all solutions to* (4.6) *with small initial conditions* $v(\xi, 0)$ *remain small for all time* $t > 0$. *The solution is said to be linearly unstable if there exists a solution with small initial condition that do not stay small in time.*

We now illustrate this process on the KdV equation (1.48), which, when written using the notation introduced in (4.1), corresponds to

$$\mathcal{P}(\partial_x) = -\partial_x^3, \text{ and } \mathcal{N}(u) = -6uu_x. \tag{4.7}$$

We look at the KdV written in the traveling variables in (1.49), which we reproduce here

$$u_t - cu_\xi + u_{\xi\xi\xi} + 6uu_\xi = 0. \tag{4.8}$$

We substitute on the LHS of (4.8) the expression $u = u_0(\xi) + v(t, \xi)$, where we assume $u = u_0(\xi)$ to be a traveling wave solution such as the pulse solution given in (1.51). We find

$$\begin{aligned} u_t - cu_\xi + u_{\xi\xi\xi} + 6uu_\xi &= u_0''' - cu_0' + 6u_0u_0' \\ &+ v_t - cv_\xi + v_{\xi\xi\xi} + 6u_0v_\xi + 6u_0'v + 6vv_\xi, \end{aligned} \tag{4.9}$$

where the prime notation u_0' denotes the derivative of u_0 with respect to ξ (remember that u_0 only depends on ξ). Because u_0 solves the traveling wave version of the KdV (4.8), we have that

$$u_0''' - cu_0' + 6u_0u_0' = 0. \tag{4.10}$$

The first three terms of the RHS of (4.9) thus add up to zero. Dropping the term $6vv_\xi$ that is nonlinear in v and equating the right side of (4.9) to zero, we find the KdV linearized about the traveling wave solution u_0 written as

$$v_t = cv_\xi - v_{\xi\xi\xi} - 6u_0v_\xi - 6u_0'v. \tag{4.11}$$

Notice that this is a linear equation for the perturbation v with coefficients determined in terms of the traveling wave solution u_0. Notice also that we have used an "intuitive" approach consisting of substituting the perturbed solution in the equations and dropping any terms that are of degree more than one in the perturbation and its derivatives. Alternatively, one could have used equation (4.6) to obtain the linearization.

Equation (4.11) captures the evolution of a small perturbation. However, equation (4.11) is linear and it was obtained after the nonlinear term was removed. It thus represents a linear approximation of the evolution equation for a perturbation to u_0. As such, the study of (4.11) is related to what we call "linear stability or instability" as outlined in Definition 4.1. In this context, we say that the solution to the KdV (1.48) is linearly unstable if there is a solution to (4.11) that initially is small but grows indefinitely in time. Indeed, such a solution would represent a perturbation to the solution that would grow in time. In general, it can be challenging to solve (4.11) to establish linear stability or to find a solution growing in time. Instead, we are going to look at a special kind of perturbations of the form

$$v(t, \xi) = w(\xi)e^{\lambda t}, \tag{4.12}$$

where we have assumed separation of variables and an exponential time dependence. For reasons that will become clear below, we will refer to the constant λ as the spectral parameter. With such an assumption, equation (4.11) becomes

$$\lambda w = -w''' + cw' - 6u_0 w' - 6u_0' w, \tag{4.13}$$

where the prime $'$ denotes derivative with respect to ξ. While making this assumption is obviously restrictive, the advantage of (4.13) over (4.11) is that it is an ordinary differential equation and ordinary differential equations are usually easier to solve than partial differential equations. Furthermore, the behavior of perturbations of the form (4.12) in time are easy to understand. For example, if w solves (4.13) for a certain real positive value of λ, then (4.12) represents a solution to the linearized KdV (4.11) that grows exponentially in time.

As we mentioned before, what we want to define as an instability is a "small" perturbation that grows indefinitely in time, which, in the case of (4.12) and (4.13) means exponential growth. The question now is what does one mean by small for a function $w(\xi)$ defined for $\xi \in \mathbb{R}$. Modeling a physical system on an infinite domain (all of $\mathbb{R}$) is of course an abstraction and one would want any perturbation to have no contribution as $\xi \to \pm\infty$. It turns out that a very convenient way to ensure that the perturbation w decays to zero as $\xi \to \pm\infty$ is to implement that condition

$$\int_{-\infty}^{\infty} |w|^2 d\xi < \infty, \tag{4.14}$$

that is, we want the integral of the square of the norm of w on $\mathbb{R}$ to exist. When it does, we use it to define the "norm" $\|w\|$ of the function w, that is,

$$\|w\| \equiv \sqrt{\int_{-\infty}^{\infty} |w|^2 d\xi}. \tag{4.15}$$

The formula above extends the idea of a norm of a vector to functions. In (4.15), the integrand is $|w|$ to ensure the norm is a real non-negative number, even in the case where w is a complex valued function. We need to consider the possibility that w be complex since, for example, if λ in (4.13) is a complex number, then the space of solutions to (4.13) will be complex as well. We denote the set of functions satisfying condition (4.14) by $L^2(\mathbb{R})$ and call it the space of "square-integrable" functions. It is a vector space defined in the following way

$$L^2(\mathbb{R}) = \left\{ w : \mathbb{R} \to \mathbb{C} \,\middle|\, \int_{-\infty}^{\infty} |w|^2 d\xi < \infty \right\}, \tag{4.16}$$

that is, $L^2(\mathbb{R})$ consists of the measurable complex-valued function defined on $\mathbb{R}$ that are "square-integrable," meaning that the integral in (4.15) exists. The set $L^2(\mathbb{R})$ is a vector space because the sum of two functions in that set is also a function satisfying (4.15). This can be shown by using the fact that if w_1 and w_2 are two complex valued functions (or two complex numbers), then

$$|w_1 + w_2|^2 \leq 2 \left(|w_1|^2 + |w_2|^2 \right).$$

In effect, by introducing the condition (4.16), one ensures that the perturbations w:

(i) decay fast enough to zero as $\xi \to \pm\infty$ and

(ii) if there is a singularity at finite values of ξ, it does not diverge "too badly".

For example, the following functions

$$f_1 = \text{sech}(\xi), \quad f_2 = e^{-|\xi|}, \quad f_3 = \frac{e^{-|\xi|}}{|\xi|^{1/4}}, \quad f_4 = \frac{1}{1+\xi^2},$$

are part of $L^2(\mathbb{R})$ because their corresponding norms as defined in (4.15) exist. In the first two cases, there are no singular points at finite values of ξ. Furthermore, they both converge to zero exponentially fast as $\xi \to \pm\infty$. The condition in (4.16) is satisfied and the norms of f_1 and f_2 can be computed from equation (4.15) as

$$\|f_1\| = \sqrt{2}, \quad \|f_2\| = 1.$$

In the case of f_3, the decay to zero as $\xi \to \pm\infty$ is also exponential. Although there is a singular point at $\xi = 0$, the integral exists and we have

$$\|f_3\| = (2\pi)^{1/4}.$$

In other words, the singular behavior ($f_3 \sim 1/|\xi|^{1/4}$ around $\xi = 0$) is not "too bad".

Finally, in the last case, f_4 converges algebraically to zero as $\xi \to \pm\infty$. However, it does decay fast enough for the integral in (4.15) to exist and we have

$$\|f_4\| = \sqrt{\frac{\pi}{2}}.$$

The following functions are examples that are not in $L^2(\mathbb{R})$:

$$f_1 = \tanh(\xi), \quad f_2 = \frac{\text{sech}(\xi)}{|\xi|}, \quad f_3 = \frac{e^{-|\xi|}}{\sqrt{|\xi|}}, \quad f_4 = \frac{1}{\sqrt{1+|\xi|}}.$$

The norm of each of those functions as defined in (4.15) does not exist because the corresponding integrals diverge. The first one does not decay to zero at infinity, thus the integral of $|f_1|^2$ on $\mathbb{R}$ diverges. The second and third functions have a singular behavior at $\xi = 0$ that is not integrable. Around the value $\xi = 0$, the function $|f_2|^2$ behaves as $1/\xi^2$, while $|f_3|^2$ behaves as $1/|\xi|$. The function f_4 has no singularities and it does converge to zero as $\xi \to \pm\infty$. However, f_4 does not decay to zero fast enough and the integral (4.15) diverges.

Now that we have introduced the concept of the norm of a function on $\mathbb{R}$, we can state a precise definition of linear stability as follows.

Definition 4.2 *We say that the traveling wave solution $u = u_0(\xi)$ to (4.2) is* linearly *stable if, for every $\epsilon > 0$, there exists a $\delta > 0$ such that, any solution v to (4.6) with initial conditions satisfying $\|v(\xi, 0)\| < \epsilon$ exists for all time $t > 0$ and $\|v(t, \xi)\| < \delta$. Otherwise, it is said to be* linearly unstable.

Note that in Definition 4.2, we use the norm $L^2(\mathbb{R})$-norm (4.15). However, the reader should be aware of the fact that other function spaces are often used when establishing stability, in which case a different norm is used. In this book, we will restrict ourselves to $L^2(\mathbb{R})$. Note also that in order to simplify the notation, we restrict the definition to the scalar case. However, one can easily apply those statements to the case where we are dealing with a system of two equations or more by viewing u as a vector valued function $\mathbf{u} \in \mathbb{R}^n$. In that case, the $L^2(\mathbb{R})$ norm of a vector-valued function $\mathbf{w} \in \mathbb{R}^n$ is defined by

$$\|\mathbf{w}\| \equiv \sqrt{\int_{-\infty}^{\infty} \mathbf{w} \cdot \overline{\mathbf{w}} \, d\xi}.$$

where $\mathbf{w} \cdot \overline{\mathbf{w}}$ denotes the inner product between $\mathbf{w}$ and its complex conjugate.

The next step to continue the study of linear stability is to write (4.13) in the following way

$$\mathcal{L}w = \lambda w, \tag{4.17}$$

where, in our specific case, $\mathcal{L}$ is the following differential operator

$$\mathcal{L} = -\partial_\xi^3 + c\partial_\xi - 6u_0\partial_\xi - 6u_{0\xi}. \tag{4.18}$$

Equation (4.17) is reminiscent of an eigenvalue problem for a matrix. Indeed, if A is a n by n matrix, then the eigenvalue problem for A reads

$$A\mathbf{x} = \lambda\mathbf{x},$$

where $\mathbf{x}$ is an n-dimensional vector. For a differential operator such as $\mathcal{L}$ defined in (4.18), we will still call (4.17) eigenvalue problem except that we will only be interested in functions v that are in $L^2(\mathbb{R})$. To be precise, we introduce the following definition.

Definition 4.3 *We call (4.17) an eigenvalue problem on $L^2(\mathbb{R})$ and say that a value $\lambda = \lambda_0$ is an eigenvalue if equation (4.17) has a solution $v \in L^2(\mathbb{R})$ for that particular value of λ.*

For example, the reader is encouraged to compute the derivative u_0' of the solution given in (1.51) and show that it satisfies

$$\mathcal{L}u_0' = 0, \tag{4.19}$$

where $\mathcal{L}$ is given in (4.18). Then check that $w = u_0'$ satisfies the condition (4.14), by computing the integral explicitly. This shows that $\lambda = 0$ is an eigenvalue of $\mathcal{L}$ given in (4.18), with eigenvector $w = u_0'$. Actually, (4.19) can be obtained in a different way by simply taking the derivative of the traveling wave equation for the KdV given in (4.10). After doing so, one precisely equation (4.19) with $\mathcal{L}$ given in (4.18). Thus, equation (4.19) shows that $\lambda = 0$ is an eigenvalue for the operator $\mathcal{L}$ defined in (4.18) since it corresponds to (4.17) in the case where λ is zero. In the general case, equation (4.19) is obtained by taking the derivative of the traveling wave equation given in (4.3). We thus obtain the following result.

Theorem 4.1 *Consider the eigenvalue problem* (4.17) *obtained from making the substitution* (4.12) *into the linearization* (4.6) *of* (4.1) *about a traveling wave solution* $u(t, x) = u_0(\xi)$. *Then* $\lambda = 0$ *is an eigenvalue for* (4.17), *that is* (4.19) *is satisfied.*

The result described by Theorem 4.1 is attributed to the translational invariance property of the starting partial differential equation (4.2) that if it admits a solution $u = \widetilde{u}(t, \xi)$ (not necessarily a traveling wave solution) then $u = \widetilde{u}(t, \xi + \xi_0)$ also is a solution for any real value of ξ_0. This is due to the fact that the partial differential equation (4.2) does not depend explicitly on the variable ξ (or that (4.1) does not depend explicitly on the variable x). As a consequence, when differentiating (4.3), one obtains (4.19). For example, consider the KdV (4.8). It is easily checked using the chain rule that when performing the translation $\xi \rightarrow \xi + \xi_0$ on any solution, then one obtains a translate of the solution of the KdV. That is the reason why the KdV traveling wave solution (1.51) is a solution for any value of the translation parameter ξ_0. Such a translation symmetry does not apply, for example, in a partial differential equation such as

$$u_t = x u_{xx} \tag{4.20}$$

because it depends explicitly on x. Equation (4.20) is not an example that falls in the class of equations described by (4.1).

The eigenvalue problem (4.17) is obtained from the linearization by the assumption (4.12) on a solution of the linearization. Thus, if (4.17) has an $L^2(\mathbb{R})$ solution for a given value of λ with positive real part (in other words, the operator $\mathcal{L}$ has an eigenvalue with positive real part), then the linearization has a solution initially small in $L^2(\mathbb{R})$ that grows exponentially in time. Therefore we have the following preliminary result that relates the eigenvalue problem (4.17) to linear stability.

Theorem 4.2 *Consider the linearization* (4.6) *of equation* (4.2) *about a traveling wave solution. Assume the eigenvalue problem* (4.17) *arising from the linearization under the assumption* (4.12) *has an eigenvalue with positive real part. The traveling wave solution is then linearly unstable.*

Consider now Example (1.41), the Nagumo model. Written in the notation defined in (4.1), we have

$$P(\partial_x) = \partial_x^2, \text{ and } N(u) = u(1-u)(u-a). \qquad (4.21)$$

We consider a traveling wave solution $u = u_0(\xi)$, $\xi = x - ct$, and obtain the linearization as

$$v_t = v_{\xi\xi} + cv_\xi + \left(2(a+1)u_0 - 3u_0^2 - a\right)v. \qquad (4.22)$$

The reader should obtain the linearization above either using equation (4.6) or by substituting $u = u_0(\xi) + v(t, \xi)$ into (1.41), and, like done in the KdV case, reject the terms that are nonlinear in v and its derivatives. The eigenvalue problem (4.17) is obtained by making the substitution (4.12) into (4.22) and it corresponds to the following differential operator

$$\mathcal{L} = \partial_\xi^2 + c\partial_\xi + \left(2(a+1)u_0 - 3u_0^2 - a\right). \qquad (4.23)$$

Our next example is the high Lewis number combustion system (2.40). The procedure described above extends to systems, and in this case, if we write in the notation defined in (4.1), we have

$$P(\partial_x) = \begin{pmatrix} 1 & 0 \\ 0 & 0 \end{pmatrix} \partial_x^2 \text{ and } N(u, y) = \begin{pmatrix} y\Omega(u) \\ -\beta y\Omega(u) \end{pmatrix}. \qquad (4.24)$$

We consider a two-component traveling solution $(u, y) = (u_0(\xi), y_0(\xi))$ of system (2.40) and we denote by v_1 the perturbation to u and by v_2 the perturbation of y, that is,

$$u = u_0 + v_1,$$
$$y = y_0 + v_2.$$

We find the following linearization about that two-component traveling solution (u_0, y_0)

$$
\begin{aligned}
v_{1t} &= v_{1\xi\xi} + c v_{1\xi} + v_2 \,\Omega(u_0) + y_0 \Omega'(u_0)\, v_1, \\
v_{2t} &= c v_{2\xi} - \beta \left(y_0 \Omega'(u_0) v_1 + \Omega(u_0)\, v_2 \right).
\end{aligned}
\tag{4.25}
$$

Note that we have used the following Taylor expansion for the term involving Ω

$$\Omega(u_0 + v_1) = \Omega(u_0) + \Omega'(u_0)\, v_1 + \mathcal{O}(v_1^2).$$

Since the traveling waves considered here are all positive, we use the expression $\Omega(u) = e^{-1/u}$ given in (2.33) and rewrite (4.25) as

$$
\begin{aligned}
v_{1t} &= v_{1\xi\xi} + c w_{1\xi} + e^{-1/u_0}\left(\frac{y_0}{u_0^2} v_1 + v_2 \right), \\
v_{2t} &= c v_{2\xi} - \beta e^{-1/u_0}\left(\frac{y_0}{u_0^2} v_1 + v_2 \right).
\end{aligned}
\tag{4.26}
$$

The eigenvalue problem of the form

$$\mathcal{L}\mathbf{w} = \lambda \mathbf{w}, \quad \mathbf{w} = (w_1, w_2)^T$$

is obtained by the substitution

$$v_i = w_i(\xi) e^{\lambda t}, \quad i = 1, 2, \tag{4.27}$$

into (4.26). In effect, the substitution (4.27) replaces the derivatives with respect to time t in (4.26) by the multiplication by the spectral parameter λ. The corresponding linear operator takes the following form

$$
\mathcal{L} = \begin{pmatrix} 1 & 0 \\ 0 & 0 \end{pmatrix} \partial_\xi^2 + \begin{pmatrix} c & 0 \\ 0 & c \end{pmatrix} \partial_\xi + e^{-1/u_0} \begin{pmatrix} \frac{y_0}{u_0^2} & 1 \\ -\beta \frac{y_0}{u_0^2} & -\beta \end{pmatrix}
\tag{4.28}
$$

that is to be applied to vectors of the form

$$\mathbf{w} \equiv \begin{pmatrix} w_1 \\ w_2 \end{pmatrix}.$$

4.2.2 Spectrum and spectral stability

In this section, we aim at describing the spectrum of a differential operator arising in the linearization of nonlinear equations of the form (4.1). Examples of such operators are obtained in (4.18), (4.23), and (4.28). As we discussed in the previous section, eigenvalues of a differential operators are important in that they play a role in the stability properties of a solution. The notion of eigenvalue extends to the concept of spectrum, which we discuss below. We consider a differential operators $\mathcal{L}$ defined on $L^2(\mathbb{R})$ that arises from the linearization of equations of the form (4.2) about a traveling pulse or fronts through the process described in the previous section. For such an operator, we define the "resolvent set" $\rho(\mathcal{L})$ as follows (see [61, Section 2.2.4] and [89, Proposition 2.7])

$$\rho(\mathcal{L}) \equiv \left\{ \lambda \in \mathbb{C} \middle| \mathcal{L} - \lambda I \text{ is a bijection between } \mathcal{D}(\mathcal{L}) \text{ and } L^2(\mathbb{R}) \right\},$$

where I is the identity operator, i.e. $Iv = v$ for all $v \in L^2(\mathbb{R})$. Furthermore, $\mathcal{D}(\mathcal{L})$ denotes the domain of the operator $\mathcal{L}$ on $L^2(\mathbb{R})$, that is, the functions of $L^2(\mathbb{R})$ on which the operator $\mathcal{L}$ can be applied. For example, not all functions in $L^2(\mathbb{R})$ are differentiable and thus applying $\mathcal{L}$ on such functions does not make sense.

Any value of $\lambda \in \rho(\mathcal{L})$ is called a regular value. We then define the spectrum as the set of values of λ that are not in $\rho(\mathcal{L})$. We denote the spectrum of $\mathcal{L}$ as $\sigma(\mathcal{L})$, that is,

$$\sigma(\mathcal{L}) \equiv \mathbb{C} \backslash \rho(\mathcal{L}).$$

In order to understand why this extends the elementary notion of eigenvalue, consider the case where the operator is a matrix A, that is, $\mathcal{L} = A$. In that case, the resolvent set can be characterized in terms of the determinant since a matrix is invertible only if its determinant is nonzero:

$$\rho(A) = \left\{ \lambda \in \mathbb{C} \middle| \det(A - \lambda I) \neq 0 \right\}.$$

By our definition above, the spectrum of A consists of values of λ not in $\rho(A)$. Thus the spectrum characterized by the equation $\det(A - \lambda I) = 0$. This precisely corresponds to the notion of eigenvalue for the matrix A.

In order to compute the spectrum $\sigma(\mathcal{L})$ of an operator $\mathcal{L}$, we first mention that it can be divided into two subsets: the point spectrum $\sigma_p(\mathcal{L})$ and the essential spectrum $\sigma_e(\mathcal{L})$ [44,87]. We will define the point spectrum below but, for now, let us mention that it includes the "discrete spectrum" that corresponds to the set isolated eigenvalues of $\mathcal{L}$ of finite multiplicity. Remember that we defined the eigenvalues to be the values of λ for which the eigenvalue problem (4.17) has a solution that is in $L^2(\mathbb{R})$. By isolated we mean that there is a neighborhood around that value in which it is the only eigenvalue. The multiplicity refers to the dimension of the space generated by a set of eigenvectors and generalized eigenvectors (see Section 2.2.4 of [61] and Section 3.3 of [87] for a precise definition). The rest of the spectrum is called the essential spectrum and it is denoted by $\sigma_e(\mathcal{L})$. We thus have

$$\sigma(\mathcal{L}) = \sigma_p(\mathcal{L}) \cup \sigma_e(\mathcal{L}), \quad \sigma_p(\mathcal{L}) \cap \sigma_e(\mathcal{L}) = \emptyset.$$

In order to properly define the spectrum, we first rewrite the eigenvalue problem (4.17) as a linear first-order system of the form

$$\mathbf{X}' = \mathcal{A}(\xi, \lambda)\mathbf{X}, \tag{4.29}$$

where $\mathbf{X}$ is an n-dimensional vector-valued function of ξ, $\mathcal{A}$ is an n by n matrix-valued function of ξ and λ, and n corresponds to the order of (4.17). As before, we use the prime $'$ in (4.29) to denote the derivative with respect to ξ.

We now rewrite the three specific cases presented in the previous section in the form (4.29). Consider the linear operator associated to the KdV equation as given in (4.18). The corresponding eigenvalue problem reads

$$w''' = (c - 6u_0)w' - (\lambda + 6u_{0\xi})w. \tag{4.30}$$

One can turn the equation above into a 3-dimensional linear first-order system of the form (4.29) by defining the variables

$$\mathbf{X} \equiv \begin{pmatrix} x_1 \\ x_2 \\ x_3 \end{pmatrix}, \quad \text{with } x_1 \equiv w, \quad x_2 \equiv w', \quad x_3 \equiv w''.$$

We thus have that $x_1' = w' = x_2$, $x_2' = w'' = x_3$ and $x_3' = w'''$. In effect, the substitution above enables us to forego of the need to take any derivative of order more than one and the eigenvalue problem (4.30) can be written in the form (4.29) with the matrix A being given by

$$\mathcal{A} \equiv \begin{pmatrix} 0 & 1 & 0 \\ 0 & 0 & 1 \\ -\lambda - 6u_{0\xi} & c - 6u_0 & 0 \end{pmatrix}. \tag{4.31}$$

The operator (4.23) corresponding to the Nagumo model (1.41) is second order. Similarly to what was done to obtain (4.31), the eigenvalue problem corresponding to the linear operator (4.23) can be cast into a 2-dimensional linear dynamical system of the form (4.29). The matrix $\mathcal{A}$ defining that system is given by

$$\mathcal{A} \equiv \begin{pmatrix} 0 & 1 \\ a + 3u_0^2 - 2(a+1)u_0 + \lambda & -c \end{pmatrix}. \tag{4.32}$$

Finally, consider the linear operator (4.28) corresponding to the linearization of system (2.40) about a traveling wave $(u_0(\xi), y_0(\xi))$. The eigenvalue problem (4.17) can be turned into a system of the form (4.29) by introducing the vector

$$\mathbf{X} = \begin{pmatrix} x_1 \\ x_2 \\ x_3 \end{pmatrix} \equiv \begin{pmatrix} w_1 \\ w_1' \\ w_2 \end{pmatrix}.$$

The corresponding matrix takes the form

$$\mathcal{A} = \begin{pmatrix} 0 & 1 & 0 \\ \lambda - \frac{y_0}{u_0^2} e^{-1/u_0} & -c & -e^{-1/u_0} \\ \frac{\beta y_0}{c u_0^2} e^{-1/u_0} & 0 & \frac{\lambda}{c} + \frac{\beta}{c} e^{-1/u_0} \end{pmatrix}. \tag{4.33}$$

It is suggested that the reader obtains the matrices (4.32) and (4.33) on their own to ensure the understanding of the process.

Remember that an eigenvalue is a value of λ for which (4.17) has a solution that is in $L^2(\mathbb{R})$. We thus are interested in solutions to (4.29) that decay to zero fast enough as $\xi \to \pm\infty$ so that the integral condition (4.14) holds. In order to determine the asymptotic behavior (i.e. as $\xi \to \pm\infty$) of the solution of (4.29), we introduce the two matrices $\mathcal{A}^{+\infty}$ and $\mathcal{A}^{-\infty}$

$$\mathcal{A}^{\pm\infty} \equiv \lim_{\xi \to \pm\infty} \mathcal{A}(\xi, \lambda). \tag{4.34}$$

As it turns out, the solutions to (4.29) "behave like" the solutions of the asymptotic systems

$$X' = \mathcal{A}^{\pm\infty}(\lambda)X, \tag{4.35}$$

as $\xi \to \pm\infty$. Since the matrices $\mathcal{A}^{\pm\infty}$ are constant matrices (i.e. do not depend on ξ) in the cases where u_0 is a pulse or a front, the solutions to (4.35) can be obtained simply by an eigenvalue computation. For example, if σ_+ is an eigenvalue of $\mathcal{A}^{+\infty}$ with corresponding eigenvector v_+, then claiming that solutions of (4.29) "behave like" solutions to (4.35) means that there is a solution X^+ to (4.29) that asymptotically looks like

$$X^+ \sim v_+ e^{\sigma_+ \xi} \text{ as } \xi \to \infty.$$

While the statement above is admittedly vague and not mathematically sound, it has the advantage of providing an intuitive idea of how the solutions to (4.29) behave for large values of ξ. More precisely, the statement can be written as follows: there is a solution X^+ to (4.29) that satisfies

$$\lim_{\xi \to \infty} X^+ e^{-\sigma_+ \xi} = v_+.$$

The implication for our purpose is this: if $\mathcal{A}^{+\infty}$ has n_+ distinct eigenvalues with negative real parts, then system (4.29) has n_+ linearly independent solutions that converge to zero as $\xi \to \infty$, while if $\mathcal{A}^{-\infty}$ has n_- distinct eigenvalues with negative real parts, then system (4.29) has n_- linearly independent solutions that converge to zero as $\xi \to -\infty$.

We consider the system determined by the matrix (4.31) arising from the KdV equation and the traveling wave pulse solution given in (1.51). Since $\lim_{\xi \to \pm\infty} u_0 = 0$, we have from the definitions of the asymptotic matrices given in (4.34) that

$$\mathcal{A}^{+\infty}(\lambda) = \mathcal{A}^{-\infty}(\lambda) = \begin{pmatrix} 0 & 1 & 0 \\ 0 & 0 & 1 \\ -\lambda & c & 0 \end{pmatrix}. \tag{4.36}$$

The formulas for the eigenvalues of those matrices are rather complicated. However, since we really are only interested in the signs of the real parts, the exact expressions are not important. Rather, we obtain the characteristic equation for the eigenvalues

$$r^3 - cr + \lambda = 0. \tag{4.37}$$

For example, in the case $c = 7$ and $\lambda = 6$, the roots are 1, 2, and -3. While in the case $c = 2$ and $\lambda = 4$, the roots are -2, $1+i$, and $1-i$. In both cases, two of the eigenvalues have positive real part while the other one has a negative real part. As a consequence, the linear system specified by the matrix given in (4.31) has two linearly independent solutions that converge to zero as $\xi \to -\infty$ (due to the two eigenvalues with positive real parts) and only one as $\xi \to \infty$ (due to the eigenvalue with negative real part). Note that there are also diverging solutions (two as $\xi \to \infty$ and one as $\xi \to -\infty$). However, since we are looking for solutions that are in $L^2(\mathbb{R})$, we are only interested in solutions that converge at $+\infty$ or $-\infty$. We will use the characteristic equation (4.37) in the next section to compute the spectrum of the operator defined in (4.18).

We now consider the system determined by the matrix (4.32) arising from the linearization of the Nagumo model (1.41) about a traveling wave solution $u = u_0(\xi)$. We will consider the traveling fronts in the case $0 < a < 1/2$ with conditions

$$\lim_{\xi \to -\infty} u_0(\xi) = 0, \quad \lim_{\xi \to +\infty} u_0(\xi) = 1 \tag{4.38}$$

as specified in (1.42). Under the limits above, we have

$$\mathcal{A}^{+\infty}(\lambda) = \begin{pmatrix} 0 & 1 \\ 1 - a + \lambda & -c \end{pmatrix} \text{ and } \mathcal{A}^{-\infty}(\lambda) = \begin{pmatrix} 0 & 1 \\ a + \lambda & -c \end{pmatrix}.$$

The characteristic equations for the eigenvalues of those two matrices are

$$\begin{aligned} r^2 + cr - (1 - a + \lambda) &= 0 \text{ for } \mathcal{A}^{+\infty}, \\ r^2 + cr - (a + \lambda) &= 0 \text{ for } \mathcal{A}^{-\infty}. \end{aligned} \tag{4.39}$$

Since $c > 0$ (see Section 1.3 for a general discussion about the sign of c) and since $0 < a < 1/2$, it is a consequence of the quadratic formula that, in the case where $\mathrm{Re}(\lambda) > 0$, each of those two equation will have a root with a positive real part and another root with a negative one. The dynamical system determined by the matrix (4.32) thus have one linearly independent solution converging to zero as $\xi \to \infty$ and another converging one as $\xi \to -\infty$.

As a third example, we consider the system determined by the matrix (4.33) arising from the linearization of system (2.40) about a traveling wave solution $(u, y) = (u_0(\xi), y_0(\xi))$. Given the following limiting values

$$\begin{aligned} (u_0, y_0) &\to (0, 1) \text{ as } \xi \to +\infty \text{ and} \\ (u_0, y_0) &\to (u_B, 0) \text{ as } \xi \to -\infty, \quad u_B = 1/\beta, \end{aligned} \tag{4.40}$$

we find the following asymptotic matrices

$$\mathcal{A}^{+\infty}(\lambda) = \begin{pmatrix} 0 & 1 & 0 \\ \lambda & -c & 0 \\ 0 & 0 & \lambda/c \end{pmatrix},$$

$$\mathcal{A}^{-\infty}(\lambda) = \begin{pmatrix} 0 & 1 & 0 \\ \lambda & -c & -e^{-\beta} \\ 0 & 0 & (\lambda + \beta e^{-\beta})/c \end{pmatrix}. \tag{4.41}$$

The characteristic equation for the eigenvalues of those matrices are

$$(cr - \lambda)(r^2 + cr - \lambda) = 0 \text{ for } \mathcal{A}^{+\infty},$$
$$(cr - \lambda - \beta e^{-\beta})(r^2 + cr - \lambda) = 0 \text{ for } \mathcal{A}^{-\infty}. \tag{4.42}$$

Since c and β are both positive, we find from (4.42) that, in the case where $\mathrm{Re}(\lambda) > 0$, $\mathcal{A}^{+\infty}$ has only one eigenvalue with negative real part given by

$$-\frac{c}{2} - \frac{1}{2}\sqrt{c^2 + 4\lambda}, \tag{4.43}$$

while $\mathcal{A}^{-\infty}$ has two eigenvalues with positive real part given by

$$\frac{\lambda + \beta e^{-\beta}}{c} \text{ and } -\frac{c}{2} + \frac{1}{2}\sqrt{c^2 + 4\lambda}. \tag{4.44}$$

The dynamical system determined by the matrix (4.33) thus has one solution converging to zero as $\xi \to \infty$ and two linearly independent solutions converging to zero as $\xi \to -\infty$.

We are now ready to proceed and explain how the spectrum of a given operator $\mathcal{L}$ can be defined and can be computed. We use the notation of [87] and define the "Morse indices" $i_\pm(\lambda)$ of $\mathcal{A}^{\pm\infty}$ as

$i_\pm(\lambda) \equiv$ the number of eigenvalues of $\mathcal{A}^{\pm\infty}$ with positive real parts.

We define the point spectrum $\sigma_p(\mathcal{L})$ of $\mathcal{L}$ as follows.

Definition 4.4 *We say that an eigenvalue λ of $\mathcal{L}$ is in the point spectrum $\sigma_p(\mathcal{L})$ of $\mathcal{L}$ if the matrices (4.34) are hyperbolic and $i_+(\lambda) = i_-(\lambda)$.*

We say that a matrix is hyperbolic if it has no purely imaginary or zero eigenvalues. As mentioned before, the point spectrum does include the discrete spectrum, that is, the set of isolated eigenvalues of finite multiplicity. In practice, we will only be interested in the location of that part of the point spectrum. See Remark 3.2 of [87], which discusses the difference between the set $\sigma_p(\mathcal{L})$ and what we call here the discrete spectrum, which the author of [87] denotes Σ_{pt} and $\tilde{\Sigma}_{pt}$, respectively. Notice that by restricting the point spectrum to the eigenvalues for which $i_+(\lambda) = i_-(\lambda)$ in Definition 4.4, we effectively ask that the dimension of the space of solutions to (4.29) that go to zero as $\xi \to +\infty$ and the dimension of the space of solutions to (4.29) that go to zero as $\xi \to -\infty$ add up to the total dimension of the space, n. If the sum corresponding to a value of λ is more than n, then that value of λ is automatically an eigenvalue because the spaces have to intersect.

A happy consequence of Definition 4.4 is that, since the essential spectrum of $\mathcal{L}$ is the complement of the point spectrum, we have the following theorem:

Theorem 4.3 *Consider the linear operator $\mathcal{L}$ obtained from the linearization of a partial differential equation of the form (4.1) about a traveling wave pulse or front solution. Consider also the matrix $\mathcal{A}(\xi, \lambda)$ obtained by rewriting the eigenvalue problem (4.17) on $L^2(\mathbb{R})$ as a first-order linear system of the form (4.29). Then the essential spectrum of $\mathcal{L}$ corresponds to the values of λ for which one of the asymptotic matrices (4.34) is not hyperbolic or for which $i_+(\lambda) \neq i_-(\lambda)$.*

The difference $i_-(\lambda) - i_+(\lambda)$ is called the Fredholm index. While the reader might not have the background necessary to fully understand what the different spectra represent, we hope that the several examples we will work out below will help to build an intuitive idea. Also, let us say this: the general strategy will be to compute the essential spectrum using Theorem 4.3, and if it will turn out that it is restricted to the left side of the complex plane, then it will be sufficient, for our purpose, to look for elements of the discrete spectrum on the right side of the complex plane. If It does not, i.e. if there is an essential spectrum on the right side of the plane, then some other type of analysis may be needed to understand the type of instability the wave has.

In the discussion below, we use the term "spatial eigenvalues" for the eigenvalues of the asymptotic matrices $\mathcal{A}^{+\infty}$ or $\mathcal{A}^{-\infty}$. Along a trajectory in the complex plane, the real part of a spatial eigenvalue can change sign only if the values of λ cross one of the curves in the set defined by

$$B \equiv \{\lambda \in \mathbb{C} |\ \text{a spatial eigenvalue is purely imaginary or is zero}\}.$$
(4.45)

This is because for one of the Morse indices to change value, a real part of one of the spatial eigenvalues must change sign. Thus inside any connected region that has no intersection with the set B defined above, the Morse indices stay constant and can be determined by computing it at any point of the region.

Consider the example of KdV (1.48) and the matrix (4.31). We find the values of λ for which the matrices are not hyperbolic by assuming that at least one of the spatial eigenvalues is purely imaginary. Substituting $r = i\sigma$, $\sigma \in \mathbb{R}$, in the characteristic equation (4.37) for the matrix $\mathcal{A}^{\pm\infty}$ (in the case of a pulse, the asymptotic matrices are equal), we find

$$\lambda = i(\sigma^3 - c\sigma). \qquad (4.46)$$

Since σ is a free real parameter, equation (4.46) is telling us that the asymptotic matrices have a purely imaginary eigenvalue (i.e. they are not hyperbolic matrices) for any value of λ on the imaginary axis. Thus the value of the Morse indices $i_{\pm}(\lambda)$ can only change if the value of λ is made to cross the imaginary axis in the λ complex plane. Both indices stay constant in the regions of the complex plane on the right or the left side of the complex plane. It thus suffices to compute their value at one point in each of these regions. In the case under study here, the solution is a pulse and both asymptotic matrices are the same. Therefore the condition $i_{-}(\lambda) = i_{+}(\lambda)$ is satisfied everywhere and the essential spectrum is restricted to the imaginary axis. One way to determine the values of the Morse indices on each half complex plane, is to consider the characteristic equation (4.37) for real values of λ with large absolute values. In such a case, the roots of (4.37) approach the values of r given by

$$r^3 = -\lambda. \qquad (4.47)$$

In the case where $\lambda > 0$, this can be seen by making the substitution $r = \sqrt[3]{\lambda}\rho$ into (4.37), and dividing the equation by λ to obtain

$$\rho^3 - \frac{c}{\lambda^{2/3}}\rho + 1 = 0.$$

Taking the limit $\lambda \to \infty$, one gets

$$\rho^3 + 1 = 0,$$

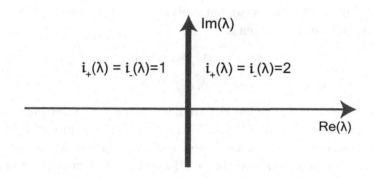

Figure 4.3 Graph of the essential spectrum of the operator (4.18) arising from the linearization of the KdV equation (1.48) about the traveling pulse solution $u = u_0(\xi)$ given in (1.51). The essential spectrum consists of the imaginary axis. The given values of $i_\pm(\lambda)$ mean that the linear system determined by the matrix (4.31) with $\mathrm{Re}(\lambda) > 0$ has one solutions converging to zero at $+\infty$ and two at $-\infty$. For $\mathrm{Re}(\lambda) < 0$, it has two solution converging to zero at $+\infty$ and one at $-\infty$.

which gives (4.47) after substituting $\rho = r/\sqrt[3]{\lambda}$. For real positive values of λ, equation (4.47) has one real negative solution and two complex conjugate solutions with positive real parts. We thus have that $i_+(\lambda) = i_-(\lambda) = 2$ when $\mathrm{Re}(\lambda) > 0$. In a similar fashion, we find $i_+(\lambda) = i_-(\lambda) = 1$ when $\mathrm{Re}(\lambda) < 0$ (see Figure 4.3). Thus, for any value of λ on the right side of the complex plane, the linear dynamical system (4.29) with matrix (4.31) has two linearly independent solutions converging to zero as $\xi \to -\infty$ and one converging to zero when $\xi \to \infty$, while the opposite holds true for values of λ on the left side of the complex plane.

Now consider the system determined by the matrix (4.32) arising from the linearization of the Nagumo equation (1.41) about a traveling wave solution $u = u_0(\xi)$ satisfying the conditions (4.38) with $0 < a < 1/2$. We compute the set B defined in (4.45) like in the previous example by substituting $r = i\sigma$ into the characteristic equations (4.39). We write λ as a complex number with its real and imaginary part $\lambda = \lambda_r + i\lambda_i$, and we substitute in (4.39). Solving the real and imaginary parts of the obtained equations, we find

$$\begin{aligned} \lambda_i &= c\sigma, \quad \lambda_r = -\sigma^2 + a - 1 \\ \lambda_i &= c\sigma, \quad \lambda_r = -\sigma^2 - a. \end{aligned} \tag{4.48}$$

Each line in (4.48) defines two parabolas on the right side of the complex plane, which can be written as

$$\lambda_r = -\lambda_i^2/c + a - 1,$$
$$\lambda_r = -\lambda_i^2/c - a. \tag{4.49}$$

Those two parabolas divide the complex plane into three regions. To determine the values of $i_-(\lambda)$ and $i_+(\lambda)$, we choose values of λ in each of those three regions. For large positive real values of λ, the zeroes of both characteristic equations in (4.39) approach the values of r defined by

$$r^2 = \lambda. \tag{4.50}$$

Equation (4.50) is obtained in a similar way as was (4.47). This implies that each characteristic equation in (4.39) has a root with positive real part and a root with negative real part in the region on the right side of the parabolas defined in (4.49). Thus, in that region, $i_-(\lambda) = i_+(\lambda) = 1$. In the region between the two parabolas, we pick the value $\lambda = -1/2$, since $a - 1 < -1/2 < -a$ for $0 < a < 1/2$. In that case, it follows from the quadratic formulas (and the fact that $c > 0$) that both roots of the second equation in (4.39) have negative real part and we have $i_-(\lambda) = 0$ and $i_+(\lambda) = 1$, and thus the region is part of the essential spectrum. Finally, in the region on the left side of both parabolas, we pick any real value $\lambda < -1/2$ to find that the equations in (4.39) each have two roots with negative real parts and we have $i_-(\lambda) = i_+(\lambda) = 0$. The essential spectrum is thus restricted to the region in between the two parabolas as illustrated in Figure 4.4.

Consider now the third example of the system determined by the matrix (4.33) arising from the linearization of system (2.40) about a traveling wave solution $(u, w_2) = (u_0(\xi), y_0(\xi))$ with boundary conditions (4.40). As in the previous examples, we find the set B defined in (4.45) by substituting $r = i\sigma$, with σ real in the characteristic equations (4.42). We find three curves define by the following:

$$\lambda_i = c\sigma, \quad \lambda_r = 0,$$
$$\lambda_i = c\sigma, \quad \lambda_r = -\beta e^{-\beta}, \tag{4.51}$$
$$\lambda_i = c\sigma, \quad \lambda_r = -\sigma^2.$$

The first of the curves defined above is the imaginary axis, the second is a vertical line on the left side of the complex plane, and the last

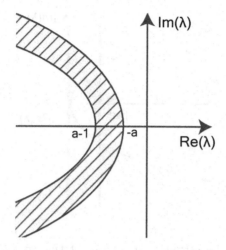

Figure 4.4 Graph of the essential spectrum of the operator (4.23) arising from the linearization of the Nagumo equation (1.41) about a traveling wave solution $u = u_0(\xi)$ satisfying the conditions (4.38) with $0 < a < 1/2$. The corresponding set B as defined in (4.45) corresponds to two parabolas restricted to the left side of the complex plane. The region in between the two parabolas is filled with essential spectrum.

one defines a parabola with a vertex at the origin. The curves defined in (4.51) divide the complex plane into seven regions. The part of the complex plane on the right side of the imaginary axis is easily found not to contain essential spectrum. Indeed, for large λ real, the roots of both characteristic equations given in (4.42) approach the values determined by

$$(cr - \lambda)(r^2 - \lambda) = 0,$$

which has two positive and one negative roots. As a consequence, each equation in (4.42) have two solutions with positive real parts for values of λ on the right side of the complex plane (i.e. with $\mathrm{Re}(\lambda) > 0$). For such λ's, we have $i_-(\lambda) = i_+(\lambda) = 2$ and there is no essential spectrum on the right of the imaginary axis. The essential spectrum is illustrated in Figure 4.5. We leave it to the reader to compute the Morse indices to conclude that, in addition to the curves defined in (4.51), the shaded regions in Figure 4.5 are what constitute the essential spectrum.

Now that we established how the essential spectrum can be computed, we need to discuss the point spectrum. The point spectrum is

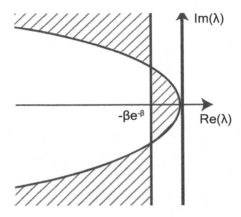

Figure 4.5 Graph of the essential spectrum of the operator (4.28) arising from the linearization system (2.40) about a traveling wave solution $(u, w_2) = (u_0(\xi), y_0(\xi))$ satisfying the conditions (4.40). The corresponding set B as defined in (4.45) corresponds to two vertical line (the imaginary axis and the vertical line $\text{Re}(\lambda) = -\beta e^{-\beta}$) and one parabola restricted to the left side of the complex plane but with vertex located at the origin. The rest of the essential spectrum corresponds to the shaded region on the graph.

determined by finding all possible values λ outside the essential spectrum for which the eigenvalue problem (4.17) has a solution in $L^2(\mathbb{R})$. While this seems to be a straightforward task, it turns out that, in general, it is quite challenging to do analytically. In the next section, we present an analytical method (energy estimate computations) and a numerical tool (Evans function) whose goal is to approximatively find the location of elements of the point spectrum. In practice, as we will explain later, we will only be interested in determining the discrete spectrum (isolated eigenvalues) located on the right side of the complex plane.

To finish this section, we present the definition of spectral stability. We have already discussed the fact that if the eigenvalue problem (4.17) has an eigenvalue on the right side of the complex plane, then the corresponding traveling wave solution is linearly unstable in the sense of Definition 4.2. As we will discuss later, it turns out that, not only eigenvalues, but any part of the spectrum that is on the right side of the complex plane can cause instability. This motivates the following definition.

Definition 4.5 *We say that the traveling wave solution* $u = u_0(\xi)$ *to* (4.2) *is* spectrally *stable if the spectrum (essential and point) of the linear operator* $\mathcal{L}$ *arising from the linearization* (4.6) *is restricted to the set* $\mathrm{Re}(\lambda) < 0$, *except for an eigenvalue at* $\lambda = 0$ *due to Theorem 4.1.*

4.3 LOCATION OF THE POINT SPECTRUM

In this section, we present two methods, which can be used together to locate the point spectrum. The first one is analytical and is called "Spectral Energy Estimates". It aims at finding a finite region of the complex plane in which elements of the discrete spectrum must be located. The second one is numerical and is called the "Evans function" method. It consists in numerically constructing an object (the Evans function) defined in the complex plane values that has zeros at the values of λ that are in the discrete spectrum. The Evans function is used to numerically locate the discrete spectrum in the finite region determined by the Spectral Energy Estimates method.

The reader should be aware that there are other methods to numerically compute the spectrum of a differential operator. For example, one can discretize the differential equation (4.17), turning the problem of determining the spectrum of an operator to the problem of finding the eigenvalues of a large matrix. Chapter 7 of [11] describes how this can be done using spectral methods for the discretization.

4.3.1 Spectral Energy Estimates

The starting point in this section will be the eigenvalue problem (4.17). Since we are only interested in spectral stability, we are looking for eigenvalues such that $\mathrm{Re}(\lambda) \geq 0$, that is, eigenvalues that are on the right side of the complex plane. Indeed, the eigenvalues on the left side of the complex plane, while interesting to discuss, have no implication on the spectral stability as defined in Definition 4.5. What we will be doing in this section, is to find a bound on the modulus $|\lambda|$ under the condition that $\mathrm{Re}(\lambda) \geq 0$. We will thus trap the eigenvalues with positive real parts inside a semicircle on the right side of the complex plane.

The general idea of Spectral Energy Estimates is to multiply the eigenvalue problem (4.17) by the complex conjugate $\overline{w}$ and integrate over $\mathbb{R}$ to obtain

$$\int_{-\infty}^{\infty} (\mathcal{L}w)\,\overline{w}\,\mathrm{d}\xi = \lambda \int_{-\infty}^{\infty} |w|^2 \mathrm{d}\xi, \tag{4.52}$$

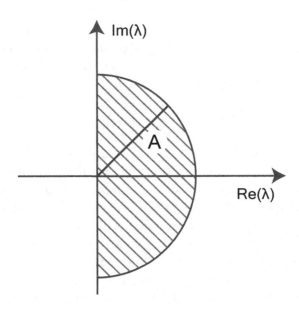

Figure 4.6 Graph of the region defined by the inequality (4.54) under the assumption that $\text{Re}(\lambda) \geq 0$.

where we used the fact that $w\bar{w} = |w|^2$. Then, through the use of various inequalities, we aim at obtaining that the left side satisfies an inequality where the quantity $\|w\|^2 = \int_{-\infty}^{\infty} |w|^2 \mathrm{d}\xi$ can be factored out, that is,

$$\left| \int_{-\infty}^{\infty} (\mathcal{L}w)\,\bar{w}\,\mathrm{d}\xi \right| \leq A\|w\|^2, \tag{4.53}$$

for some real constant A. Once (4.53) is established, one obtains from (4.52) the following inequality on the norm of λ

$$|\lambda| \leq A. \tag{4.54}$$

Under the assumption that $\text{Re}(\lambda) \geq 0$, we thus have that the eigenvalue is trapped in the right half of the circle centered at the origin and with radius A, as illustrated in Figure 4.6.

The computations necessary to obtain (4.53) will mostly rely on five inequalities, which we present here. The first one is Young's inequality [104]. If $a \geq 0$ and $b \geq 0$ are nonnegative real numbers and if $p > 1$ and $q > 1$ are real numbers such that $\frac{1}{p} + \frac{1}{q} = 1$, then

$$ab \leq \frac{a^p}{p} + \frac{b^q}{q}. \tag{4.55}$$

Note that in the particular case $p = q = 2$, Young's inequality is a direct consequence of the fact that

$$(a - b)^2 \geq 0.$$

The second one is the triangle inequality for complex numbers: if $z \in \mathbb{C}$, then

$$|z| \leq |\operatorname{Re}(z)| + |\operatorname{Im}(z)|. \tag{4.56}$$

The third one is called the Cauchy-Schwarz inequality (Theorem 7 of [42]). If the functions f and g are both in $L^2(\mathbb{R})$, then one has the inequality

$$\left| \int_{-\infty}^{\infty} f \bar{g} \, d\xi \right|^2 \leq \int_{-\infty}^{\infty} |f|^2 d\xi \int_{-\infty}^{\infty} |g|^2 d\xi. \tag{4.57}$$

The fourth inequality that we will often use can be seen as a triangle inequality for integrals. If f is integrable on $\mathbb{R}$ then we have

$$\left| \int_{-\infty}^{\infty} f d\xi \right| \leq \int_{-\infty}^{\infty} |f| d\xi. \tag{4.58}$$

Our final inequality follows from standard results taught in calculus courses. If g is integrable on $\mathbb{R}$ and f is a bounded function on $\mathbb{R}$, then

$$\int_{-\infty}^{\infty} |fg| d\xi \leq \sup(|f|) \int_{-\infty}^{\infty} |g| d\xi, \tag{4.59}$$

where $\sup(|f|)$ denotes the supremum value of $|f|$ on $\mathbb{R}$.

We start our presentation of Spectral Energy Estimates by working on the following "toy" example of an eigenvalue problem:

$$w'' + w' + \Psi(\xi)w = \lambda w, \tag{4.60}$$

where $\Psi(\xi)$ is some real-valued function on $\mathbb{R}$ that we assume to be bounded. We think of (4.60) as an eigenvalue problem on $L^2(\mathbb{R})$ not necessarily thinking of it as originating from a spectral stability problem. We will use this simple example to illustrate the basic method when it comes to find a bound on any eigenvalue in the right side of the complex plane. We multiply (4.60) by $\bar{w}$ and integrate to obtain

$$\int_{-\infty}^{\infty} w'' w d\xi + \int_{-\infty}^{\infty} w' \bar{w} d\xi + \int_{-\infty}^{\infty} \Psi(\xi)|w|^2 d\xi = \lambda \int_{-\infty}^{\infty} |w|^2 d\xi. \tag{4.61}$$

The first two terms on the left side are the "problematic" ones in the sense that we cannot factor the quantity $\|w\|^2 = \int_{-\infty}^{\infty} |w|^2 d\xi$. However, we use integration by parts on the first term to obtain

$$\int_{-\infty}^{\infty} w'' \overline{w} d\xi = w' \overline{w} \Big|_{-\infty}^{\infty} - \int_{-\infty}^{\infty} w' \overline{w}' d\xi.$$

Since we are assuming λ is an eigenvalue, the corresponding solution w is in $L^2(\mathbb{R})$ and thus decays to zero as $\xi \to \pm\infty$. We thus have

$$\int_{-\infty}^{\infty} w'' \overline{w} d\xi = -\int_{-\infty}^{\infty} |w'|^2 d\xi, \tag{4.62}$$

and (4.61) becomes

$$-\int_{-\infty}^{\infty} |w'|^2 d\xi + \int_{-\infty}^{\infty} w' \overline{w} d\xi + \int_{-\infty}^{\infty} \Psi(\xi)|w|^2 d\xi = \lambda \int_{-\infty}^{\infty} |w|^2 d\xi. \tag{4.63}$$

We find a bound on both the real part and imaginary part of λ. We do this by taking the real part and imaginary part of (4.63) and then using the inequalities mentioned before to find bounds on $\mathsf{Re}(\lambda)$ and $\mathsf{Im}(\lambda)$. We start by taking the real part of (4.63) to obtain

$$\mathsf{Re}(\lambda) \int_{-\infty}^{\infty} |w|^2 d\xi = -\int_{-\infty}^{\infty} |w'|^2 d\xi + \int_{-\infty}^{\infty} \Psi(\xi)|w|^2 d\xi. \tag{4.64}$$

To obtain (4.64), we have used several facts. First, we have that the quantities

$$\int_{-\infty}^{\infty} |w|^2 d\xi, \int_{-\infty}^{\infty} |w'|^2 d\xi, \text{ and } \int_{-\infty}^{\infty} \Psi(\xi)|w|^2 d\xi$$

are real. Second, the quantity $\int_{-\infty}^{\infty} w' \overline{w} d\xi$ is purely imaginary. To see this, we apply integration by parts and obtain

$$\int_{-\infty}^{\infty} w' \overline{w} d\xi = |w|^2 \Big|_{-\infty}^{\infty} - \int_{-\infty}^{\infty} w \overline{w}' d\xi = -\int_{-\infty}^{\infty} \overline{w}' w d\xi, \tag{4.65}$$

where we have used the fact that w is in $L^2(\mathbb{R})$, and like in (4.62), the term evaluated at $\pm\infty$ is zero. In (4.65), if we define the integral on the left side to be

$$I \equiv \int_{-\infty}^{\infty} w' \overline{w} d\xi,$$

then the right side of (4.65) is equal to $-\overline{I}$. Thus, $I = -\overline{I}$ and the integral $\int_{-\infty}^{\infty} w'\overline{w}\mathrm{d}\xi$ is purely imaginary. We can now take the imaginary part of equation (4.63) to obtain

$$\mathsf{Im}(\lambda) \int_{-\infty}^{\infty} |w|^2 \mathrm{d}\xi = -i \int_{-\infty}^{\infty} w'\overline{w}\mathrm{d}\xi, \qquad (4.66)$$

where we have multiplied the purely imaginary quantity $\int_{-\infty}^{\infty} w'\overline{w}\mathrm{d}\xi$ by $-i$ to obtain its imaginary part. We take the modulus on both sides of (4.66) and the resulting eqution is given by

$$|\,\mathsf{Im}(\lambda)\,| \int_{-\infty}^{\infty} |w|^2 \mathrm{d}\xi = \left| \int_{-\infty}^{\infty} w'w\mathrm{d}\xi \right|. \qquad (4.67)$$

Using the inequality (4.58) for the right side of (4.67) implies

$$|\,\mathsf{Im}(\lambda)\,| \int_{-\infty}^{\infty} |w|^2 \mathrm{d}\xi \leq \int_{-\infty}^{\infty} |w'||w|\mathrm{d}\xi. \qquad (4.68)$$

Note that we used the fact that $|\overline{z}| = |z|$ for any complex number z. Adding (4.64) and (4.68), we obtain the following inequality on the sum of the real and imaginary parts of λ:

$$(\mathsf{Re}(\lambda) + |\,\mathsf{Im}(\lambda)\,|) \int_{-\infty}^{\infty} |w|^2 \mathrm{d}\xi \leq - \int_{-\infty}^{\infty} |w'|^2 \mathrm{d}\xi$$
$$+ \int_{-\infty}^{\infty} |w'||w|\mathrm{d}\xi + \int_{-\infty}^{\infty} \Psi(\xi)|w|^2 \mathrm{d}\xi. \qquad (4.69)$$

There is an issue caused by the presence of w' on the right side. The last term can be dealt with using inequality (4.59) as we will see below. For now, we apply Young's inequality (4.55) in the case $p = q = 2$, $a = \sqrt{2}|w'|$, and $b = |w|/\sqrt{2}$ to obtain

$$|w'||w| \leq |w'|^2 + \frac{|w|^2}{4}. \qquad (4.70)$$

The reasoning in using Young's inequality in the form above is that once it is substituted into the second term of the right side of (4.69), then the terms involving w' disappear. Indeed, using (4.70) into (4.69), one finds

$$(\mathsf{Re}(\lambda) + |\,\mathsf{Im}(\lambda)\,|) \int_{-\infty}^{\infty} |w|^2 \mathrm{d}\xi \leq \frac{1}{4} \int_{-\infty}^{\infty} |w|^2 \mathrm{d}\xi$$
$$+ \int_{-\infty}^{\infty} \Psi(\xi)|w|^2 \mathrm{d}\xi. \qquad (4.71)$$

To deal with the last term, we apply inequality (4.59) and obtain

$$\int_{-\infty}^{\infty} \Psi(\xi)|w|^2 d\xi \leq \sup(|\Psi|) \int_{-\infty}^{\infty} |w|^2 d\xi. \tag{4.72}$$

As explained before in (4.53), it is important to be able to pull out the factor $\|w\|^2 = \int_{-\infty}^{\infty} |w|^2 d\xi$. Substituting (4.72) into (4.71), dividing by $\|w\|^2$, and then use inequality (4.56) on λ, we find

$$|\lambda| \leq \frac{1}{4} + \sup(|\Psi|). \tag{4.73}$$

Thus what we found is that any eigenvalue to (4.60) with non-negative real part must satisfy inequality (4.73).

We now turn to the case of the Nagumo equation (1.41) and the corresponding linear operator (4.23) arising from the linearization about a traveling wave solution. The eigenvalue equation (4.17) for this particular operator reads

$$\lambda w = w'' + cw' + \left(2(a+1)u_0 - 3u_0^2 - a\right) w. \tag{4.74}$$

The front solution $u = u_0(\xi)$ can be any of the two solutions satisfying the conditions (1.42) in the case $0 < a < 1/2$ or $a > 1/2$. As we did for the previous case, we multiply (4.74) by $\overline{w}$ and integrate to get

$$\lambda \int_{-\infty}^{\infty} |w|^2 d\xi = -\int_{-\infty}^{\infty} |w'|^2 d\xi + c\int_{-\infty}^{\infty} w'\overline{w} d\xi - a\int_{-\infty}^{\infty} |w|^2 d\xi$$
$$+ 2(a+1)\int_{-\infty}^{\infty} u_0|w|^2 d\xi - 3\int_{-\infty}^{\infty} u_0^2|w|^2 d\xi,$$

where the first term was obtained by integration by parts as in (4.62). Like in the previous example, we take the real and imaginary parts and find

$$\mathsf{Re}(\lambda) \int_{-\infty}^{\infty} |w|^2 d\xi = -\int_{-\infty}^{\infty} |w'|^2 d\xi - a\int_{-\infty}^{\infty} |w|^2 d\xi$$
$$+ 2(a+1)\int_{-\infty}^{\infty} u_0|w|^2 d\xi - 3\int_{-\infty}^{\infty} u_0^2|w|^2 d\xi, \tag{4.75}$$

$$|\,\mathsf{Im}(\lambda)\,| \int_{-\infty}^{\infty} |w|^2 d\xi \leq c\int_{-\infty}^{\infty} |w'||w| d\xi.$$

Taking into account the coefficient c on the right side of the second equation in (4.75), we apply Young's inequality (4.55) in the case $p = q = 2$, $a = \sqrt{2}|w'|$, and $b = c|w|/\sqrt{2}$ to obtain

$$c|w'||w| \le |w'|^2 + \frac{c^2|w|^2}{4}. \tag{4.76}$$

As in the previous example, the term involving w' disappears when adding the two equations in (4.75), and we obtain

$$(\text{Re}(\lambda) + |\,\text{Im}(\lambda)\,|) \int_{-\infty}^{\infty} |w|^2 d\xi \le \frac{c^2}{4} \int_{-\infty}^{\infty} |w|^2 d\xi - a \int_{-\infty}^{\infty} |w|^2 d\xi$$
$$+ 2(a+1) \int_{-\infty}^{\infty} u_0 |w|^2 d\xi - 3 \int_{-\infty}^{\infty} u_0^2 |w|^2 d\xi. \tag{4.77}$$

Notice that the second and fourth terms on the right side of (4.77) are negative, we can thus forego them and conclude that inequality (4.77) implies

$$(\text{Re}(\lambda) + |\,\text{Im}(\lambda)\,|) \int_{-\infty}^{\infty} |w|^2 d\xi \le \frac{c^2}{4} \int_{-\infty}^{\infty} |w|^2 d\xi$$
$$+ 2(a+1) \int_{-\infty}^{\infty} u_0 |w|^2 d\xi.$$

We use inequality (4.59) on the second term of the right side and divide by $\|w\|^2$ to obtain

$$|\lambda| \le \frac{c^2}{4} + 2(a+1) \sup(|u_0|).$$

As a last example of Spectral Energy Estimates, we consider the eigenvalue problem equation (4.17) associated to the operator defined in (4.28). Recall that this operator arises the linearization of the combustion model (2.42) about the traveling wave $(u, y) = (u_0(\xi), y_0(\xi))$ satisfying the conditions (2.39). From (4.17) and (4.28), the eigenvalue problem reads

$$\lambda w_1 = w_1'' + c w_1' + e^{-1/u_0} w_2 + \frac{y_0}{u_0^2} e^{-1/u_0} w_1,$$
$$\lambda w_2 = c w_2' - \beta e^{-1/u_0} w_2 - \beta \frac{y_0}{u_0^2} e^{-1/u_0} w_1. \tag{4.78}$$

Since (4.78) is a system, the strategy must be adapted from the scalar case. What really makes this example different from the previous ones is the fact that the second equation is first order. We adapt the strategy used in [97, Section 3] for a case that generalizes (4.78) to more than one space dimension. We first solve the second equation to obtain an inequality for $|w_2|$. The strategy for the first equation will be similar to what we have done so far in this section, except that the inequality for $|w_2|$ will be used to obtain a right hand side in terms of w_1 only.

We rewrite the second equation in (4.78) as

$$w_2' - \frac{1}{c}\left(\lambda + \beta e^{-1/u_0}\right)w_2 = \beta\frac{y_0}{cu_0^2}e^{-1/u_0}w_1. \tag{4.79}$$

Equation (4.79) can be regarded as a linear first-order non-homogeneous differential equation for w_2. To solve it, we multiply it by the integrating factor

$$\mu \equiv \exp\left(\frac{1}{c}\left(\beta\int_\xi^\infty e^{-1/u_0}d\hat{\xi} - \lambda\xi\right)\right), \tag{4.80}$$

where the improper integral is known to exist since, from (4.40), $u_0 \to 0^+$ as $\xi \to \infty$. The equation (4.79) can then be written as

$$\frac{d}{d\xi}\left(\mu w_2\right) = \beta\mu\frac{y_0}{cu_0^2}e^{-1/u_0}w_1.$$

We obtain w_2 by integration

$$w_2 = -\frac{\beta}{c\mu}\int_\xi^\infty\left(\mu\frac{y_0}{u_0^2}e^{-1/u_0}w_1\right)d\hat{\xi}, \tag{4.81}$$

where everything in the integrand is taken to be dependent on $\hat{\xi}$. In principle, one should write $\mu(\hat{\xi})$, $y_0(\hat{\xi})$, $u_0(\hat{\xi})$, and $w_1(\hat{\xi})$ in the integrand above, while using $\mu(\xi)$ for the μ outside the integral. However, for simplicity of notation, we ask the reader to assume that the integrands in (4.80) and (4.81) are all taken to depend on $\hat{\xi}$. From (4.81), we now use inequality (4.58) and the Cauchy-Schwarz inequality (4.57) to obtain the following:

$$|w_2| \le \frac{\beta}{c|\mu|}\sqrt{\int_\xi^\infty\left(|\mu|^2\frac{y_0^2}{u_0^4}e^{-2/u_0}\right)d\hat{\xi}}\sqrt{\int_{-\infty}^\infty|w_1|^2d\hat{\xi}}$$

$$\le \frac{\beta}{c\tilde{\mu}|e^{-\lambda\xi/c}|}\sqrt{|e^{-2\lambda\xi/c}|\int_\xi^\infty\left(\tilde{\mu}^2\frac{y_0^2}{u_0^4}e^{-2/u_0}\right)d\hat{\xi}}\sqrt{\int_{-\infty}^\infty|w_1|^2d\hat{\xi}}$$

$$\le \frac{\beta}{c\tilde{\mu}}\sqrt{\int_\xi^\infty\left(\tilde{\mu}^2\frac{y_0^2}{u_0^4}e^{-2/u_0}\right)d\hat{\xi}}\sqrt{\int_{-\infty}^\infty|w_1|^2d\hat{\xi}}.$$

$$\tag{4.82}$$

where μ is given in (4.80) and

$$\tilde{\mu} \equiv \mu e^{\lambda \xi / c} = \exp\left(\frac{\beta}{c} \int_\xi^\infty e^{-1/u_0} \mathrm{d}\hat{\xi}\right). \tag{4.83}$$

On the second line of (4.82), we are able to pull out the function $|e^{-2\lambda \xi / c}|$ from the integral using inequality (4.59), and the fact that

$$\sup_{\hat{\xi} \in [\xi, \infty)} \left(\left|e^{-2\lambda \hat{\xi} / c}\right|\right) = \left|e^{-2\lambda \xi / c}\right|,$$

since we assume $\mathrm{Re}(\lambda) \geq 0$. Remember that λ is an eigenvalue, thus w_1 is in $L^2(\mathbb{R})$, implying that the integrals in (4.82) are all convergent.

We now multiply the first equation of (4.78) by $\overline{w}_1$, integrate, separate the real and imaginary parts, and simplify to obtain

$$\mathrm{Re}(\lambda) \int_{-\infty}^\infty |w_1|^2 \mathrm{d}\xi = -\int_{-\infty}^\infty |w_1'|^2 \mathrm{d}\xi + \int_{-\infty}^\infty e^{-1/u_0} \mathrm{Re}(w_2 \overline{w}_1) \, \mathrm{d}\xi$$

$$+ \int_{-\infty}^\infty \frac{y_0}{u_0^2} e^{-1/u_0} |w_1|^2 \mathrm{d}\xi,$$

$$\mathrm{Im}(\lambda) \int_{-\infty}^\infty |w_1|^2 \mathrm{d}\xi = -ic \int_{-\infty}^\infty w_1' \overline{w}_1 \mathrm{d}\xi$$

$$+ \int_{-\infty}^\infty e^{-1/u_0} \mathrm{Im}(w_2 \overline{w}_1) \, \mathrm{d}\xi. \tag{4.84}$$

As in the previous examples, we have used integration by parts in the first equations and the fact that the integral of $w_1' \overline{w}_1$ in the second equation of (4.84) is purely imaginary. We take the absolute value of the second equation and use the triangle inequality. We obtain

$$|\mathrm{Im}(\lambda)| \int_{-\infty}^\infty |w_1|^2 \mathrm{d}\xi \leq c \left|\int_{-\infty}^\infty w_1' \overline{w}_1 \mathrm{d}\xi\right|$$

$$+ \left|\int_{-\infty}^\infty e^{-1/u_0} \mathrm{Im}(w_2 \overline{w}_1) \, \mathrm{d}\xi\right|. \tag{4.85}$$

We apply inequality (4.58) to (4.85) and get

$$|\mathrm{Im}(\lambda)| \int_{-\infty}^\infty |w_1|^2 \mathrm{d}\xi \leq c \int_{-\infty}^\infty |w_1'| |w_1| \mathrm{d}\xi$$

$$+ \int_{-\infty}^\infty e^{-1/u_0} |\mathrm{Im}(w_2 \overline{w}_1)| \, \mathrm{d}\xi. \tag{4.86}$$

As in Example (4.74), we apply Young's inequality given in (4.76) for $c|w_1'||w_1|$ and add the first equation of (4.84) and the inequality for $\mathsf{Im}(\lambda)$ obtained in (4.86) and get

$$|\lambda| \int_{-\infty}^{\infty} |w_1|^2 \mathrm{d}\xi \le \frac{c^2}{4} \int_{-\infty}^{\infty} |w_1|^2 \mathrm{d}\xi + \int_{-\infty}^{\infty} \frac{y_0}{u_0^2} e^{-1/u_0} |w_1|^2 \mathrm{d}\xi$$

$$+ \int_{-\infty}^{\infty} e^{-1/u_0} \left(\mathsf{Re}(w_2 \overline{w}_1) + |\, \mathsf{Im}(w_2 \overline{w}_1)\,| \right) \mathrm{d}\xi.$$

We use the fact that, for any complex number z, we have the inequality

$$(|\, \mathsf{Re}(z)\,| + |\, \mathsf{Im}(z)\,|) \le \sqrt{2}|z|$$

and write

$$|\lambda| \int_{-\infty}^{\infty} |w_1|^2 \mathrm{d}\xi \le \frac{c^2}{4} \int_{-\infty}^{\infty} |w_1|^2 \mathrm{d}\xi + \int_{-\infty}^{\infty} \frac{y_0}{u_0^2} e^{-1/u_0} |w_1|^2 \mathrm{d}\xi$$

$$+ \sqrt{2} \int_{-\infty}^{\infty} e^{-1/u_0} |w_2 w_1| \mathrm{d}\xi.$$

We are now ready to use inequality (4.82) for $|w_2|$

$$|\lambda| \int_{-\infty}^{\infty} |w_1|^2 \mathrm{d}\xi \le \frac{c^2}{4} \int_{-\infty}^{\infty} |w_1|^2 \mathrm{d}\xi + \int_{-\infty}^{\infty} \frac{y_0}{u_0^2} e^{-1/u_0} |w_1|^2 \mathrm{d}\xi$$

$$+ \frac{\sqrt{2}\beta}{c} \sqrt{\int_{-\infty}^{\infty} |w_1|^2 \mathrm{d}\xi} \int_{-\infty}^{\infty} \left(|w_1| \frac{e^{-1/u_0}}{\tilde{\mu}} \sqrt{\int_{\xi}^{\infty} \tilde{\mu}^2 \frac{y_0^2}{u_0^4} e^{-2/u_0} \mathrm{d}\hat{\xi}} \right) \mathrm{d}\xi,$$

where $\tilde{\mu}$ is given in (4.83). We use the Cauchy-Schwarz inequality (4.57) on the last term and the inequality (4.59) on the second term. This enables us to factor the expression $\|w_1\|^2 = \int_{-\infty}^{\infty} |w_1|^2 \mathrm{d}\xi$ on the right side and we obtain the following inequality

$$|\lambda| \le \frac{c^2}{4} + \sup\left(\frac{y_0}{u_0^2} e^{-1/u_0} \right)$$

$$+ \frac{\sqrt{2}\beta}{c} \sqrt{\int_{-\infty}^{\infty} \left(\frac{e^{-2/u_0}}{\tilde{\mu}^2} \int_{\xi}^{\infty} \tilde{\mu}^2 \frac{y_0^2}{u_0^4} e^{-2/u_0} \mathrm{d}\hat{\xi} \right) \mathrm{d}\xi}. \tag{4.87}$$

Remember that the result above concerns the traveling wave solutions to combustion model (2.42), which itself is obtained from the model

(2.32) by setting the parameter ϵ to zero. Because we considered $\epsilon = 0$, the second equation in the eigenvalue problem (4.78) is a first-order equation. The strategy to deal with that equation has been to solve it for w_2 and obtain inequality (4.82) for $|w_2|$. Now, in the case where $\epsilon \neq 0$, the eigenvalue problem reads

$$\lambda w_1 = w_1'' + c w_1' + e^{-1/u_0} w_2 + \frac{y_0}{u_0^2} e^{-1/u_0} w_1,$$

$$\lambda w_2 = \epsilon w_2'' + c w_2' - \beta e^{-1/u_0} w_2 - \beta \frac{y_0}{u_0^2} e^{-1/u_0} w_1. \tag{4.88}$$

At this point the reader should be able to derive (4.88) from linearizing (2.32) about a traveling wave $(u, y) = (u_0(\xi), y_0(\xi))$ as it is a slight generalization of the process used to obtain (4.78). Notice that the second equation in (4.88) is of second order, and as a consequence, the way by which the bound is found is more in line with the first two examples treated in this section. By this it means that the bound is obtained by multiplying the first equations in (4.88) by $\overline{w}_1$ and the second one by $\overline{w}_2$ before integrating. The process is worked out in detail in [29] and we encourage the reader to go over this computation.

Other examples of Spectral Energy Estimates can be found in [30–34, 54, 97].

4.3.2 Evans function

4.3.2.1 Definition of the Evans function

In the previous section, we have identified a bounded region of the complex plane where every element of the point spectrum with nonnegative real part is located. The Evans function is a tool that we use to find any eigenvalue in that region. While the Evans function can in principle be computed analytically, in practice, it usually turns out to be an impossible task. One however can always compute it numerically and use it to find the location of eigenvalues.

In order to define the Evans function, we use the formulation of the eigenvalue problem (4.17), that is,

$$\mathcal{L}w = \lambda w. \tag{4.89}$$

We rewrite (4.89) as a linear first-order dynamical system of the form given in (4.29), that is of the form

$$\mathbf{X}' = \mathcal{A}(\xi, \lambda)\mathbf{X}, \tag{4.90}$$

where $\mathbf{X}$ is an n-dimensional vector value function of ξ, A is an n by n matrix-valued function of ξ and λ. We assume λ not to be in the essential spectrum of $\mathcal{L}$. It follows from Theorem 4.3 that the Fredholm index $i_-(\lambda) - i_+(\lambda)$ is zero and the asymptotic matrices $\mathcal{A}^{\pm\infty}$ are hyperbolic (i.e. they do not have purely imaginary eigenvalues). Thus, if we denote by n_+ the number of eigenvalues of $\mathcal{A}^{+\infty}$ with negative real parts and by n_- the number of eigenvalues of $\mathcal{A}^{-\infty}$ with positive real parts, we have that $n_+ + n_- = n$. Consider for example the case of the system determined by the matrix (4.33) arising from the linearization of the high Lewis number combustion model (2.40) about a traveling wave solution $(u, y) = (u_0(\xi), y_0(\xi))$. For $\text{Re}(\lambda) > 0$, it was found that the asymptotic matrices given in (4.41) are such that $n_+ = 1$ and $n_- = 2$. The corresponding linear system thus has one solution converging to zero as $\xi \to \infty$, and two solutions converging to zero as $\xi \to -\infty$.

In general, since the solutions of system (4.90) behave like the solutions of the asymptotic system

$$\mathbf{X}' = \mathcal{A}^{\pm\infty}\mathbf{X}, \tag{4.91}$$

as $\xi \to \pm\infty$, system (4.90) has an n_+-dimensional space of solutions converging to zero as $\xi \to \infty$ and an n_--dimensional space of solutions converging to zero as $\xi \to -\infty$. Furthermore, two bases

$$\{\mathbf{X}_i^+\}_{i=1}^{n_+} \text{ and } \{\mathbf{X}_i^-\}_{i=1}^{n_-} \tag{4.92}$$

for the solutions converging, respectively, at $+\infty$ and $-\infty$ can be chosen to be analytic in λ in any simply-connected region in which $n_+ + n_- = n$ and in which the asymptotic matrices are hyperbolic (see [87, Section 4.1] and [63, Chapter II.4.2]). In other words, the bases in (4.92) can be chosen to be analytic in any simply-connected region outside of the essential spectrum.

We are looking for the point spectrum of $\mathcal{L}$. This means we are looking for values λ outside of the essential spectrum for which the eigenvalue problem (4.89) has a solution in $L^2(\mathbb{R})$. Such solutions converge to zero both as $\xi \to +\infty$ and $\xi \to -\infty$. This happens if the spaces spanned by the sets in (4.92) have a nonzero intersection (see Figure 4.7). To determine if the two spaces intersect, we first evaluate the solutions in (4.92) at some value of ξ. For example, we may choose $\xi = 0$. That is, we define

$$\mathbf{U}_i^+(\lambda) \equiv \mathbf{X}_i^+(\xi = 0), i = 1, \ldots, n_+, \quad \mathbf{U}_i^-(\lambda) \equiv \mathbf{X}_i^-(\xi = 0), i = 1, \ldots, n_-.$$

Do the two spaces intersect at ξ=0?

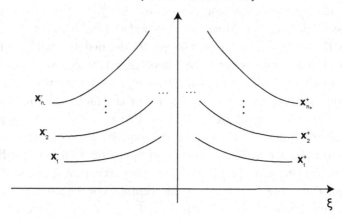

Figure 4.7 System (4.90) defines a vector space of dimension n_+ of solutions that converge to zero as $\xi \to \infty$ and another one of dimension n_- of solutions that converge to zero as $\xi \to -\infty$. A given value of λ is an eigenvalue for the corresponding operator $\mathcal{L}$ if the two spaces intersect.

Then, there is a non-trivial intersection between the two spaces spanned by the bases (4.92) if the following determinant

$$D(\lambda) \equiv \det\left(\mathbf{U}_1^+, \mathbf{U}_2^+, \dots, \mathbf{U}_{n_+}^+, \mathbf{U}_1^-, \mathbf{U}_2^-, \dots, \mathbf{U}_{n_-}^-\right) \qquad (4.93)$$

is zero. The function $D(\lambda)$ above is called the Evans function and a given value $\lambda = \lambda_0$ is an eigenvalue for the problem (4.89), if and only if

$$D(\lambda_0) = 0.$$

In fact, we have the following theorem (Theorem 4.1 of [87]).

Theorem 4.4 *Consider a simply-connected region Ω of the complex plane outside of the essential spectrum of the operator $\mathcal{L}$. For values of λ restricted to Ω, we have the following properties of the Evans function D defined in (4.93).*

- *The Evans function is analytic.*

- *$D(\lambda_0) = 0$ if and only if λ_0 is an eigenvalue of $\mathcal{L}$.*

- *The Evans function has a zero at $\lambda = 0$, that is, $D(0) = 0$.*

The region Ω is taken simply connected and outside of the essential spectrum because, in such a region, n_+ and n_- will not change value and they will add up the dimension of system (4.90), that is, $n_+ + n_- = n$. As discussed before, the Morse indices $i_\pm(\lambda)$ (and thus $n_\pm$) can only change if the values of λ cross the set B defined in (4.45). The fact that $D(0) = 0$ is a consequence of Theorem 4.1, which says that $\lambda = 0$ is an eigenvalue with eigenvector $w = u_0'$. As mentioned above, the result about analyticity is a consequence of the fact that the two bases (4.92) can be chosen analytically. Those two bases are not unique and thus the Evans function is not defined uniquely. However, if we restrict ourselves to cases where $\mathcal{L}$ is a real is an operator with real coefficients, the eigenvalue problem (4.89) has the property that if $w = w_0$ is a solution for the value $\lambda = \lambda_0$, then the complex conjugate $w = \overline{w}_0$ solves the same eigenvalue problem with $\lambda = \overline{\lambda}_0$. As a consequence, we can always choose the Evans function so that it is symmetric with respect to the real axis. That is, we have the following theorem:

Theorem 4.5 *It is possible to choose the two bases* $\{X_i^+\}_{i=1}^{n_+}$ *and* $\{X_i^-\}_{i=1}^{n_-}$ *in (4.92) so that the Evans function satisfies the following properties.*

- $D(\lambda)$ *is real whenever* $\lambda \in \mathbb{R}$.

- $D(\overline{\lambda}) = \overline{D(\lambda)}$.

Theorem 4.4 transforms the problem of finding eigenvalue of $\mathcal{L}$ into the problem of finding the zeros of an analytic function. An immediate consequence is the following. In all our examples, the right side of the complex plane will be outside the essential spectrum. We can thus take that simply connected region Ω to be the whole right half of the plane. Since the Evans function is analytic, it is either zero in all that region, or it has isolated zeros only. If we are able to show, from energy estimate computations, that there is a bound on the modulus $|\lambda|$ of any eigenvalue with positive real part, it implies that the Evans function is not zero everywhere. Thus, there can only be a discrete spectrum (i.e. isolated eigenvalues) on the right side of the complex plane. Our task is thus to locate those isolated zeros.

Given a simply-connected region of finite size with boundary given by a curve C, the strategy to determine if the Evans function D has zeros inside that region is to figure out its winding number while λ moves along the curve, that is, the number of times the graph of D

travels counterclockwise around the origin. This winding number can be obtained by computing the line integral of the logarithmic derivative of D along the curve, that is,

$$\text{winding number of } D \text{ along } C = \frac{1}{2\pi i} \oint_C \frac{D'(\lambda)}{D(\lambda)} d\lambda. \qquad (4.94)$$

By Cauchy's argument principle, the integral above gives the number of zeros (counting multiplicity) of D inside the region bounded by the curve C. Here, we are assuming D does not have any singular point inside the region. This is true by Theorem 4.4 if the region is taken outside of the essential spectrum. We are also assuming that D does not have zeros on the curve C itself. We thus have to avoid the origin $\lambda = 0$ when choosing the curve C, since $D(0) = 0$ by Theorem 4.4. The strategy is now straightforward. If one wants to find out if a finite size region of the complex plane outside the essential spectrum contains any eigenvalue, one needs to compute the integral above. And there is more. In most cases, and (as mentioned before) in all the examples we are considering, the essential spectrum will be restricted to the left side of the complex plane (that is, values of λ for which $\text{Re}(\lambda) \leq 0$) and we will only be interested in determining if there are eigenvalues that create instabilities. Hence, we will only be interested in finding the eigenvalues that are on the right side of the complex plane. Furthermore, in the cases where the technique of Spectral Energy Estimates described in Section 4.3.1 works at finding a bound, one has trapped the unstable eigenvalues in a finite region of the complex plane such as illustrated in Figure 4.6. One can thus in principle use the computation described above in (4.94) to find all of the unstable eigenvalues.

Remember that by Theorem 4.4, the Evans function has a zero at the origin ($\lambda = 0$) of the complex plane. Thus, the origin cannot be included along the curve C used in (4.94) to compute the winding number. If one wants to use a closed curve that encircles a region of the form depicted in Figure 4.6, the curve has to travel on the left side of the complex plane to avoid the point $\lambda = 0$. If the essential spectrum is on the left side of the complex plane and bounded away from the imaginary axis, then a closed curve that surrounds a region of the form of Figure 4.6 and that avoids the origin can be chosen outside the essential spectrum where Theorem 4.4 applies. For example, consider the case the case of the operator (4.23) arising from the linearization of the Nagumo equation (1.41) about a traveling wave solution $u = u_0(\xi)$ satisfying the conditions (4.38) with $0 < a < 1/2$. Figure 4.4 shows that there is a gap between the

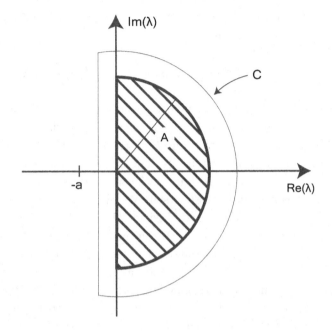

Figure 4.8 Example of a curve C that encircles a region such as the one illustrated in Figure 4.6, while remaining outside essential spectrum generated by the Nagumo equation from Figure 4.4.

imaginary axis and the essential spectrum on the left side of the complex plane. Figure 4.8 illustrates a possible choice of such a curve C that would encircle a region of the form depicted in Figure 4.6, while staying outside of the essential spectrum. In the case of the essential spectrum illustrated in Figure 4.5 corresponding to combustion fronts, finding such a curve is not possible because the imaginary axis itself is part of the essential spectrum. However, the Evans function can be extended in an analytic fashion to values of the spectral parameter λ for which the asymptotic matrices have eigenvalues that change sign, that is, when the spectral parameter enters the essential spectrum. Analytic continuation of the Evans function across the essential spectrum is addressed by the Gap Lemma [28, 62] (see also Theorem 10.1.2 of [61] and the discussions in Section 4.3 of [87] and Section 3.2 of [70]).

4.3.2.2 Gap Lemma

To present the Gap Lemma, we come back to the n-dimensional linear system (4.90). We remind the reader that outside the essential spectrum, the number n_+ of eigenvalues of $\mathcal{A}^{+\infty}$ with negative real parts and n_- of eigenvalues of $\mathcal{A}^{-\infty}$ with positive real parts are such that $n_+ + n_- = n$. We denote the corresponding n_+ eigenvalues of $\mathcal{A}^{+\infty}$ and the corresponding n_- eigenvalues of $\mathcal{A}^{-\infty}$ respectively as

$$
\begin{aligned}
\mathcal{A}^{+\infty} &: \left\{\nu_1^+, \nu_2^+, \ldots, \nu_{n_+}^+\right\}, \quad 0 \geq \operatorname{Re}(\nu_1^+) \geq \operatorname{Re}(\nu_2^+) \geq \cdots \geq \operatorname{Re}\left(\nu_{n_+}^+\right), \\
\mathcal{A}^{-\infty} &: \left\{\nu_1^-, \nu_2^-, \ldots, \nu_{n_+}^-\right\}, \quad 0 \leq \operatorname{Re}(\nu_1^-) \leq \operatorname{Re}(\nu_2^-) \leq \cdots \leq \operatorname{Re}\left(\nu_{n_-}^+\right).
\end{aligned}
\tag{4.95}
$$

As the spectral parameter λ moves into the essential spectrum, some of the eigenvalues' real part listed above can change sign. However, let us assume that the following conditions continue to hold as the values of λ venture inside the essential spectrum

$$
\begin{aligned}
\operatorname{Re}(m_+) - \operatorname{Re}(\nu_1^+) &> 0 \text{ for } \mathcal{A}^{+\infty}, \\
\operatorname{Re}(\nu_1^-) - \operatorname{Re}(m^-) &> 0 \text{ for } \mathcal{A}^{-\infty},
\end{aligned}
\tag{4.96}
$$

where

m^+ is the eigenvalue of $\mathcal{A}^{+\infty}$ not in (4.95) with the smallest real part,

m^- is the eigenvalue of $\mathcal{A}^{-\infty}$ not in (4.95) with the largest real part.

In other words, we assume that the n_+ eigenvalues of $\mathcal{A}^{+\infty}$ listed in (4.95) remain the n_+ eigenvalues of $\mathcal{A}^{+\infty}$ with the smallest real parts, even if they change sign. Likewise, we assume that the n_- eigenvalues listed in (4.95) of $\mathcal{A}^{-\infty}$ remain the n_- eigenvalues of $\mathcal{A}^{+\infty}$ with the greatest real parts, even if they change sign. In that case, we modify the definition of the Evans function by saying that the bases defined in (4.92) correspond, respectively, to the n_+ eigenvalues of $\mathcal{A}^{+\infty}$ with the smallest real parts and the n_- eigenvalues of $\mathcal{A}^{-\infty}$ with the greatest real parts through the asymptotic system (4.91) and the Evans function remains analytic in the region where the conditions in (4.96) are satisfied. Furthermore, the Evans function can even be extended analytically beyond the region defined by (4.96) and this question is addressed by the Gap Lemma, which we introduce below. We define the deviators $R_\pm$ to be

$$
R_\pm(\xi, \lambda) \equiv \mathcal{A}(\xi, \lambda) - \mathcal{A}^{\pm\infty}(\lambda),
$$

and we define the constants $\gamma_\pm > 0$ as

$$R_\pm(\xi, \lambda) = O(\exp(-\gamma_\pm|\xi|)) \text{ as } \xi \to \pm\infty, \; \gamma_\pm > 0.$$

For example, consider the case of the matrix (4.31) associated to the solution defined in (1.63). Then, in that case, the solution is a pulse and $\gamma_+ = \gamma_- = \sqrt{c}$ since that value of $\gamma_\pm$ is the smallest positive number such that the limits

$$\lim_{\xi \to \infty} \left(\mathcal{A}(\xi, \lambda) - \mathcal{A}^{\pm\infty}(\lambda) \right) e^{\gamma_\pm|\xi|}$$

exist and are finite. The reader is encouraged to check this result by using the expressions given in (1.63), (4.31), and (4.36). The Gap Lemma then can be stated as follows.

Lemma 4.1 (Gap Lemma) *The Evans function can be extended analytically, except at some possible branch cuts due to the non-analyticity of the spatial eigenvalues, to a simply-connected region of the complex plane that intersects with the essential spectrum as long as the conditions*

$$\begin{aligned}
\mathrm{Re}(m_+) - \mathrm{Re}(\nu_1^+) &> -\gamma_+, \\
\mathrm{Re}(\nu_1^-) - \mathrm{Re}(m^-) &> -\gamma_-,
\end{aligned} \tag{4.97}$$

are satisfied.

Remember that the term "spatial eigenvalues" refers to the eigenvalues of the asymptotic matrices. The differences on the right side of the inequalities in (4.97) are called the spectral gaps. The conditions in (4.97) can be seen as generalizing the conditions (4.96). However, as we will see later, the conditions (4.96) will be sufficient in order to extend the Evans function beyond the essential spectrum when performing numerical computations. This is due to the fact that the numerical scheme we use is not able to follow the sets of eigenvalues (4.95) once some of their real parts meet and cross the real parts of the other spatial eigenvalues.

4.3.2.3 Evans function computation: scalar equations

The issue of course when wanting to compute the integral in (4.94) is that, in general, computing the Evans function explicitly is not possible. There exist, however, some cases where a formula can be obtained for the Evans function. For example, in the case of the KdV equation (1.48), its linearization (4.13) for the solitary wave solution (1.51) is completely solved in [56]. Another example where the Evans function is computed explicitly is provided in [60] and it corresponds to the following equation:

$$u_t = u_{xx} - u + 2u^3. \tag{4.98}$$

The equation above admits the following stationary solution

$$u = u_0(x) \equiv \operatorname{sech}(x). \tag{4.99}$$

The time-independent solution (4.99) can be seen as a traveling wave solution with speed $c = 0$. Nevertheless, we leave it as an exercise to the reader to verify the following results about the linearization of (4.98) about the solution (4.99). The eigenvalue problem corresponds to the linear operator

$$\mathcal{L} = \partial_{xx} + (6\operatorname{sech}^2(x) - 1). \tag{4.100}$$

For λ such that $\operatorname{Re}(\lambda) > -1$, the eigenvalue problem

$$\mathcal{L}w = \lambda w$$

has the following solution that decays to zero as $x \to \infty$

$$w_+ = e^{-\sqrt{1+\lambda}x}\left[1 + \frac{\lambda}{3} + \sqrt{1+\lambda}\tanh(x) - \operatorname{sech}^2(x)\right]$$

and the following solution that decays to zero as $x \to -\infty$

$$w_- = e^{\sqrt{1+\lambda}x}\left[1 + \frac{\lambda}{3} - \sqrt{1+\lambda}\tanh(x) - \operatorname{sech}^2(x)\right].$$

Through the substitution

$$\mathbf{X} = \begin{pmatrix} x_1 \\ x_2 \end{pmatrix}, \quad x_1 \equiv w, \; x_2 \equiv w',$$

one finds that $\mathbf{X}$ satisfies the two-dimensional linear dynamical system

$$\mathbf{X}' = \mathcal{A}\mathbf{X},$$

with matrix

$$\mathcal{A} = \begin{pmatrix} 0 & 1 \\ 1 + \lambda - 6\operatorname{sech}^2(x) & 0 \end{pmatrix}.$$

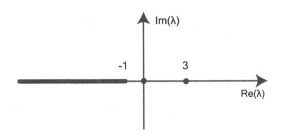

Figure 4.9 Graph of the spectrum of $\mathcal{L}$ defined in (4.100).

The Evans function can then be computed explicitly as

$$D(\lambda) = \det \begin{pmatrix} w_-(0) & w_+(0) \\ w'_-(0) & w'_+(0) \end{pmatrix} = \frac{2}{9}\lambda(3 - \lambda)\sqrt{\lambda + 1}. \qquad (4.101)$$

The Evans function thus has a zero at $\lambda = 0$ and $\lambda = 3$. The essential spectrum is easily computed using the result of Theorem 4.3 and is found to be determined by the curve

$$\lambda = -1 - \sigma^2, \quad \sigma \in \mathbb{R}. \qquad (4.102)$$

Together, equations (4.101) and (4.102) show that the spectrum of $\mathcal{L}$ defined in (4.100) consists of the values $\lambda = 0$ and $\lambda = 3$ (point spectrum) and the interval $(-\infty, -1]$ (essential spectrum), as illustrated in Figure 4.9.

Even if it is usually not possible to find a formula for the Evans function, one can usually compute numerically the Evans function and thus also the integral in (4.94). Like in any numerical computations, there are several issues when it comes to compute the Evans function (see [3] and references therein). We will circumvent them by using the MATLAB-based numerical library STABLAB that does it for us! STABLAB can be downloaded from [4]. In order to use STABLAB, one has to run the startup.m script from the main folder. Several examples are provided and each of those can be accessed in its individual folder. Each folder has a file, which contains the script to be executed. For example, in the "Burgers" folder, the file to be executed is called Burgers.m.

When a solution is known explicitly, such as the case of the Burgers equation, only two files are required to run STABLAB. The Burgers example deals with equation (1.44) when $d = 1$. In that case, the Burgers equation admits the traveling wave solution

$$u_0 = c - a \tanh(a\xi/2), \quad \xi = x - ct, \quad (4.103)$$

where a and c are free parameters, that is, equation (4.103) defines a solution of (1.44) when $d = 1$ for any real values of a and c. The folder Burgers contains the file A.m with the specification of the matrix defining the linear dynamical system (4.29) arising from the linearization of the Burgers equation about the traveling wave (4.103). The syntax to be used to enter the expression for the matrix is best learned by looking at the examples. The other file is the executable one and is called Burgers.m. Once again, the best course of action to learn how to create such a file for other cases is to look at the examples provided. However, we are going to look at some of the key elements. The first one concerns the assignment of the parameters a and c used in (4.103). The values of these are in the following two lines of the file A.m:

```
a = .5*(p.ul-p.ur);
cc = .5*(p.ul+p.ur);
```

The authors of the program decided that it is more convenient to use $p.ur$ and $p.ul$ as the free parameters, rather than a and c. The values of the variables $p.ur$ and $p.ul$ are set in the file Burgers.m and the represent the asymptotic limits of the solution u_0 at $+\infty$ ($p.ur$) and at $-\infty$ ($p.ul$). The line

```
[s,e,m,c] = emcset(s,'front',[1,1],'default');
```

calls the script emcset located in the bin folder. The first argument refers to the structure called "s" containing the various parameters of the problem. The second argument is either "front," "lopatinski," or "periodic". In the case where the solution is a front or a pulse, one chooses "front" as done above. The third argument assigns values to n_- and n_+, that is, the number $n_\pm$ of solutions going to zero as $\xi \to \pm\infty$. The last argument refers to the method to be used for the Evans function computation. STABLAB has the option of using a number of methods for that computation. By using "default" as the fourth argument, one lets the program decide which of the methods is most appropriate to be used. The lines

```
circpnts=20; imagpnts=20; innerpnts = 5; r=10;
spread=4; zerodist=10^(-2);
preimage=semicirc2(circpnts,imagpnts,innerpnts,c.
    ksteps,r,spread,zerodist,c.lambda_steps);
```

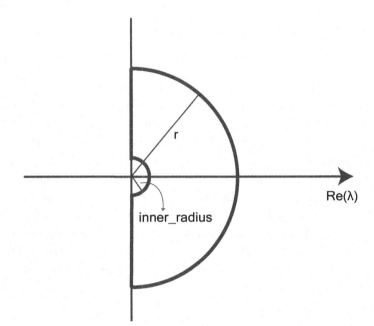

Figure 4.10 Curve along which the integral (4.94) is numerically computed when the script semicirc2 is used.

define the contour on which the integral in (4.94) is computed. Specifically, when using the script semicirc2 as above, the result is that the integral in (4.94) is computed on a semicircle whose diameter is on the imaginary axis except for a semicircle with a small inner circle skirting the origin as in Figure 4.10. The input "circpnts" is the number of points on the top half of the circle part, "imagpnts" is the number of points on the top part of the diameter along the imaginary axis, "r" is the radius of the semicircle, "spread" is a parameter that is used to spread the points on the imaginary axis so that they are more dense near the origin. This is necessary because the Evans function has a zero at the origin and thus computing the integral in (4.94) might require more points due to the fact that the integrand is expected to change more rapidly near the origin. The parameter "zerodist" (called "inner_radius" inside the semicirc2.m script) is how close along the imaginary axis the contour comes to the origin. The other inputs deal with the numerical computation of analytical bases for the eigenspaces of the asymptotic matrices $\mathcal{A}^{\pm\infty}$ defined in (4.34). As explained in Section 3(d) of [3], the numerical scheme developed in [106] is based on the method of Kato [63, p.99]. The other entries refer to that method. We give their description, even though the reader has not properly been introduced to this method of computing a basis. We believe the description of these entries will give the reader a general idea of their roles thus giving a guidance in the case problems arise that would require these entries to be changed. Our experience however has been that usually, using the values set by the creators of STABLAB has been a successful strategy. The entry "ksteps" is the number of Kato steps taken between points. If specified, "lambda_steps" is the number of contour points between Kato steps. That is, the analytic basis is not computed on the additional "lambda_steps" points unless achieving relative error tolerances of the Evans function output requires it. The line

```
s.I=12;
```

defines what the program uses as the "numerical infinity". This value is generally chosen to be a value of $|\xi|$ at which the front is close enough to its asymptotic value. For example, in the case under study in this specific program, we have $p.ul = 1$ and $p.ur = 0$. Thus $a = c = 0.5$ in (4.103) and the limiting values are

$$\lim_{\xi \to \infty} u_0(\xi) = 0 \text{ and } \lim_{\xi \to -\infty} u_0(\xi) = 1.$$

Evaluating u_0 at $\xi = \pm 12$, one finds

$$u_0(12) \approx 0.00248 \text{ and } u_0(-12) \approx 0.998.$$

As it is the case for several aspects of numerical computations, there is no set in stone rules as to how close one wants the front to be to its asymptotic values. We suggest to choose the value of $s.I$ by experimentation. If the value is too small, one might introduce an error in the Evans function computation while if it is too big, the computation time may be too long. Once more here, it is useful to experiment with several values when using the program. Figure 4.11 illustrates the graphs obtained when running the script with $s.I = 12$ and $s.I = 1$. In each case, the figures illustrate the graph of the Evans function associated to the solution (4.103) as λ varies along a curve of the form illustrated in Figure 4.10 with inner_radius = zerodist = 10^{-2} and $r = 10$. One can visually identify a problem with the figure obtained in the case $s.I = 1$ as the Evans function does not seem to be analytic. In both cases however, the

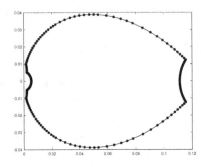 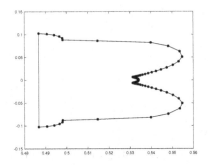

Figure 4.11 These figures are obtained when running the script Burgers.m. They illustrate the graph of the Evans function associated to the solution (4.103) as λ varies along a curve of the form illustrated in Figure 4.10 with inner_radius = zerodist = 10^{-2} and r = 10. The figure on the left was obtained the $s.I = 12$, while the one on the right with $s.I = 1$.

winding number is found to be zero as the function does not circle the origin.

Note that in addition to semicirc2.m described above, the folder bin contains the scripts semicirc.m and semibox.m which are used, respectively, when one wants to compute the integral (4.94) over a semicircle (without the inner semicircle) or over a semibox. Both scripts can be used in place of semicirc2.m in the executable script Burgers.m.

The next example concerns the Boussinesq equation

$$u_{tt} = u_{xx} - u_{xxxx} - (u^2)_{xx}, \tag{4.104}$$

with corresponding traveling wave solution

$$u_0 = \frac{1 - c^2}{2}\operatorname{sech}^2(\gamma\xi), \quad \xi = x - ct, \quad 0 < |c| < 1. \tag{4.105}$$

As far as STABLAB is concerned, the Boussinesq equation is treated in a very similar way to the Burgers equation since the traveling wave solutions for (4.104) are also known explicitly. One important detail is that the method chosen in the file is what is called "reg_adj_compound," which requires a file called Ak.m. This file is used to define an additional matrix. Rather than explaining this method and the formulas involved when creating the file Ak.m, we suggest to change the lines 20 to 25 of the executable script Boussinesq.m to a method that does not require this additional matrix in the following way:

```
% [s,e,m,c] = emcset(s,'front',[2,2],'default');
% default for Boussinesq\index{Boussinesq
    equation} is reg_reg_polar
% [s,e,m,c] = emcset(s,'front',[2,2],'
    reg_adj_polar');
% [s,e,m,c] = emcset(s,'front',[2,2],'
    adj_reg_polar');
  [s,e,m,c] = emcset(s,'front',[2,2],'
    reg_reg_polar');
%[s,e,m,c] = emcset(s,'front',[2,2],'
    reg_adj_compound');
% [s,e,m,c] = emcset(s,'front',[2,2],'
    adj_reg_compound');
```

Solution (4.105) is known to be spectrally stable when $1/2 \leq |c| < 1$ and unstable when $0 < |c| < 1/2$ (see Section 4.1 of [55] and references therein). The instability is due to a real valued positive eigenvalue (also a zero of the Evans function). Figure 2 of [55] presents a graph of the unstable eigenvalue as a function of s for $0.005 \leq c \leq 0.48$. The script Boussinesq.m is set to compute the Evans function in the case where $c = 0.4$ which, according to Figure 2 of [55], has a zero at approximatively $\lambda = 0.16$. The curve on which the Evans function is computed is defined to be a circle of radius 0.05 about $\lambda = 0.16$. The line of the script Boussinesq.m that defines that circle is given by

```
preimage = 0.16+0.05*exp(2*pi*1i*linspace(0,0.5,
    points+(points-1)*c.ksteps));
```

Note that the line above only defines the top half part of the circle as the script Boussinesq.m (and the scripts for the other examples also) is programmed to extend the curve symmetrically across the real axis by the lines

```
halfw=contour(c,s,p,m,e,preimage);
w = [halfw(1:end-1) fliplr(conj(halfw))];
```

Here, we take advantage of the fact that the Evans function itself is symmetric with respect to the real axis (see Theorem 4.5) and thus the Evans function does not need to be computed for the values of λ with negative imaginary part if the contour is symmetric with respect to the

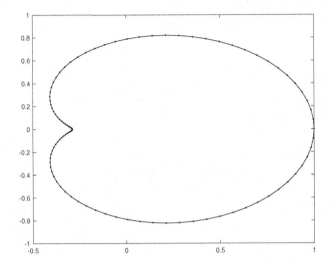

Figure 4.12 This figure is obtained when running the script Boussinesq.m. It illustrates the graph of the Evans function associated to the solution (4.105) as λ varies along the circle with radius 0.05 centered at 0.16.

real axis. When running the script Boussinesq.m with the contour as above, one obtains a winding number equal to 1 due to the eigenvalue at $\lambda = 0.16$. Figure 4.12 shows that the image of the Evans function as λ varies along the circle of radius 0.05 centered $\lambda = 0.16$ circles the origin once.

4.3.2.4 *Evans function computation: systems of equations*

The last example for the use of STABLAB for the computation of the Evans function concerns the high Lewis number combustion model (2.42) about the traveling wave $(u, y) = (u_0, y_0)$. The main differences with the previous examples is that we are dealing with a system of equations and the traveling fronts are not known explicitly. The solutions must then be computed numerically. We will use this example to illustrate how to use STABLAB when the front needs to be computed. We already have computed the fronts numerically in Section 3.4 with the MATLAB scripts located on Github at the address specified in [4] in the sub-folder HighLewis/WaveComputation. We could in principle use those computed solutions for the STABLAB Evans function computation. It turns out however, that it is simpler to just let STABLAB compute the profiles as an easy to use option. The STABLAB files for the Evans function corresponding to the combustion model are in the folder High-Lewis. The executable file for the computation of the Evans function is called HighLewis.m. In this script, we do use our programs introduced in Section 2.42 to compute the speed c of the traveling front. Although STABLAB does compute the fronts, it does not have a built-in option to determine the speed c if it is unknown. The speed is computed in the script HighLewis.m in the following line:

```
p.c = integrated_find_c(p.be);
```

In order for STABLAB to compute the front, there are two MATLAB files that need to be added that were not used for the other examples. The first one is the file F.m, that contains the right side of the nonlinear dynamical problem used to find the traveling wave solution. In the case of the combustion model (2.42), it corresponds to the right side of the system (2.45). In addition to that, the file J.m corresponds to the matrix defining the linearization of that dynamical system. In the specific case of the combustion system, the reader can use the dynamical system (2.45) and also compute the Jacobian of the right side of (2.45) in order to learn the correct syntax to be used in the files F.m and J.m. The key lines of codes in HighLewis.m for the computation of the front are the following:

```
s.n = 2; % this is the dimension of the
    profile ode
% we divide the domain in half to deal with
    the
% non-uniqueness caused by translational
    invariance
% s.side = 1 means we are solving the profile
    on the interval [0,X]
s.side=1;
s.F=@F; % F is the profile ode
s.Flinear = @J; % J is the profile ode
    Jacobian
s.UL = [1/p.be;0]; % These are the endstates
    of the profile and its derivative at x = -
    infty
s.UR = [10^(-9);1]; % These are the endstates
```

```
            of the profile and its derivative at x =
         +infty
   s.phase = [1/2/p.be; 1/2]; % this is the
         phase condition for the profile at x = 0
   s.order = [2]; % this indicates to which
         component the phase conditions is applied
   s.stats = 'on'; % this prints data and plots
         the profile as it is solved
% there are some other options you specify.
         You can look in profile_flux to see them
   [p,s] = profile_flux(p,s); % solve profile
         for first time
```

In the lines of code above, we mention that for the variable $s.UR$, we used $[10^{-9}; 1]$ rather than $[0; 1]$. This is due to the presence of the expression $e^{-1/u}$ in (2.45) that is not well-defined when $u = 0$. The reasoning behind this is explained in Section 3.4, where we perform the numerical computations of the fronts using a shooting method. The variable $s.phase$ is related to the fact that the traveling waves are translationally invariant in the sense that if a function $u_0(\xi)$ solves the traveling wave equation, then so does $u_0(\xi + \xi_0)$ for any constant ξ_0. The variable $s.phase$ is used to "fix the phase" ξ_0 of the front. We do this by fixing one of the component of the front by requesting that the front takes a given value at $\xi = 0$. The variable $s.phase$ has two components because the fronts have two as well and the condition can be applied to any one of them. By the assignment $s.order = [2]$, we ask that the second component $(u, y) = (u_0(\xi), y_0(\xi))$ be set by $s.phase$. Since $s.phase$ is set to $[1/2/p.be; 1/2]$ in the lines above, we are requesting that $y_0(0) = 1/2$. Together with the conditions (4.40) at $\pm\infty$, one then has a three-point boundary-value problem to find the front solutions numerically. As explained for example in [29], there is a way to turn this into a more conventional two-point boundary value problem restricted to either the positive $\mathbb{R}^+$ or negative $\mathbb{R}^-$ real half-line. By choosing $s.side = 1$, we choose the option where this system is solved on $\mathbb{R}^+$. We suggest that the reader makes the same choice unless something unforeseen arises.

We can now use the MATLAB scripts described above to compute the Evans on any carefully chosen contour. To choose an appropriate contour, we use the bound (4.87) found for any eigenvalue with non-negative eigenvalue the operator defined in (4.28) arising from the linearization of (2.42) about the traveling wave solutions $(u, y) = (u_0(\xi), y_0(\xi))$. The

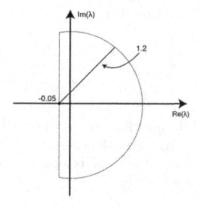

Figure 4.13 Semicircle in the complex plane defined by the lines of codes in Listing 4.1.

script BoundComputation.m is located in the folder HighLewis/Wave-Computation and it numerically computes the bound on $|\lambda|$, that is, the right side of equation (4.87) as function of β. For example, to compute the bound for $\beta = 1$, one types the following command:

```
BoundComputation(1)
```

The bound is found to be 0.9615. This means that any eigenvalue with positive real part would be trapped in a semicircle region as the one illustrated in Figure 4.6 with radius $A = 0.9615$. For the winding number computation, we thus need to choose a contour that encircles that region. We do this by the following lines in the script HighLewis.m:

Listing 4.1 Lines in HighLewis.m that define the contour of Figure 4.13

```
circpnts=30; imagpnts=30; R=1.2; spread=2;
    zerodist=0;
preimage=semicirc(circpnts,imagpnts,c.ksteps,R,
    spread,zerodist,c.lambda_steps)-0.05;
```

Listing 4.1 above specifies a semicircle whose diameter runs along the line $\text{Re}(\lambda) = -0.05$ in the λ complex plane (see Figure 4.13). As we have computed in (4.51) (see also Figure 4.5), the essential spectrum includes the imaginary axis. Thus, when defining a contour, one has to make sure the Evans function continues to be analytic in the part of the contour that "ventures" across the imaginary axis inside the essential spectrum.

From the expressions of the eigenvalues of the asymptotic matrices given in (4.43) and (4.44), we have that

$$
\begin{aligned}
\nu_1^+ &= -\frac{c}{2} - \frac{1}{2}\sqrt{c^2 + 4\,\lambda}, \\
\nu_1^- &= \frac{\lambda + \beta\,e^{-\beta}}{c} \quad \text{and} \quad \nu_2^- = -\frac{c}{2} + \frac{1}{2}\sqrt{c^2 + 4\,\lambda},
\end{aligned}
\tag{4.106}
$$

where we have used the notation introduced in (4.95) for the eigenvalues of the asymptotic matrices. In the case $\beta = 1$, the velocity is found to be $c = 0.5707$. Furthermore, the complete set of eigenvalues of the asymptotic matrices is given by

$$
\begin{aligned}
\frac{\lambda}{c} \quad \text{and} \quad -\frac{c}{2} \pm \frac{1}{2}\sqrt{c^2 + 4\,\lambda}, \quad &\text{for } \mathcal{A}^\infty, \\
\frac{\lambda + \beta\,e^{-\beta}}{c} \quad \text{and} \quad -\frac{c}{2} \pm \frac{1}{2}\sqrt{c^2 + 4\,\lambda}, \quad &\text{for } \mathcal{A}^{-\infty}.
\end{aligned}
$$

It can then be checked numerically or analytically that along the part of the semicircle of Figure 4.13 that crosses the essential spectrum (that is, on the left side of the complex plane), the eigenvalue ν_1^+ given in (4.106) stays the eigenvalue of $\mathcal{A}^\infty$ with the smallest real part, while ν_1^- and ν_2^- stay the eigenvalues of $\mathcal{A}^{-\infty}$ with the highest real parts. In other words, the conditions (4.96), thus also the less stringent conditions (4.97) stated in the Gap Lemma, is satisfied. On the right side of the complex plane, this fact is obvious since the region defined by $\mathsf{Re}(\lambda) > 0$ is outside of the essential spectrum and ν_1^+ is the only eigenvalue of $\mathcal{A}^\infty$ with the negative real part, while ν_1^- and ν_2^- are the only eigenvalues of $\mathcal{A}^{-\infty}$ with positive real part in that region. While we leave to the reader the exercise of checking that the condition (4.96) is satisfied for the part of the semicircle of Figure 4.13 on the left side of the complex plane, we included the following lines in the script HighLewis.m:

```
for cc = 1:length(preimage)
    ll=preimage(cc);
    if or(real(-p.c/2-1/2*sqrt(p.c^2+4*ll))>real(
        ll)/p.c,real(ll)<-p.c^2/4)
            disp('ALERT at + infinity')
    end;
    if or(real((ll+p.be*exp(-p.be))/p.c)<real(-p.
        c/2-1/2*sqrt(p.c^2+4*ll)),real(ll)<-p.c
        ^2/4)
```

```
        disp('ALERT at - infinity')
    end;
end
```

The lines above verify that sufficient conditions for the inequalities (4.96) are satisfied along the curve of integration. The other precautions we take due to the fact that our semicircle ventures into the essential spectrum are the following. We include the file emcset.m in the HighLewis folder. The file emcset.m is used to set the values of the STABLAB structures and it is located in [4] in the subfolder bin/bin_main. We copy the file emcset.m into the HighLewis folder. Since it is now in the folder High-Lewis, we are able to make changes to the file emcset.m that will only be used when running scripts in that folder. In the file emcset.m, we are changing the word "projection2" for the word "projection5". The scripts projection2.m and projection5.m are both used to compute projections into the eigenspaces of the asymptotic matrices. However, when the signs of the real parts of some of the eigenvalues change (this happens when venturing into the essential spectrum), projection2.m may compute projections in the wrong spaces since it relies on the signs of the eigenvalues to construct the eigenspaces. However, to compute the projections in the eigenspaces used to construct the Evans function, the script projection5.m uses the kl eigenvalues of $\mathcal{A}^{-\infty}$ with largest real part and the kr eigenvalues of $\mathcal{A}^{\infty}$ with the smallest real part. This is in contrast to the script projection2.m, which use all the eigenvalues of $\mathcal{A}^{-\infty}$ with positive real part and the eigenvalues of $\mathcal{A}^{\infty}$ with negative real part. In the case of the combustion model (2.42), we set $kl = 2$ and $kr = 1$. This enables us to venture into the essential spectrum as long as condition (4.96) is satisfied. Note that one might ask why we would not just impose the less stringent condition (4.97) stated in the Gap Lemma. The reason is that, once condition (4.96) ceases to be satisfied, then the real parts of some of the eigenvalues have met. In the case of the combustion model, that implies that either ν_1^+ given in (4.106) is not the eigenvalue of $\mathcal{A}^{\infty}$ with the smallest real part, or that ν_1^- and ν_2^- are not the eigenvalues of $\mathcal{A}^{-\infty}$ with the highest real parts. When this happens, STABLAB has no way to follow the correct eigenspaces since projection5.m will always choose the two eigenvalues of $\mathcal{A}^{-\infty}$ with the highest real parts and the one eigenvalue of $\mathcal{A}^{\infty}$ with the smallest real part.

With all the precautions described above, we run the script High-Lewis.m with $\beta = 1$ and find that if the winding number is computed along the curve illustrated in Figure 4.13, the winding number is equal

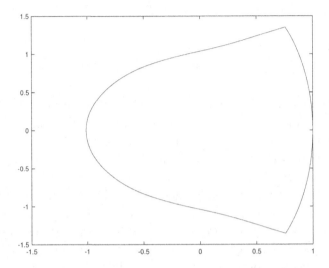

Figure 4.14 Figures obtained when running the script HighLewis.m in the case $\beta = 1$. It illustrates the graph of the Evans function associated to the traveling wave solution of the high Lewis number combustion model (2.42) as λ varies along semicircle illustrated in Figure 4.13. The winding number is found to be 1 as the graph circles the origin only once.

to 1. That is consistent with the fact that the Evans function has a zero at the origin (see Theorem 4.4) due to the translation invariance of traveling wave solutions. Figure 4.14 illustrates the graph of the Evans function for the values of λ taken along the curve depicted in Figure 4.13 and it provides a visual way to determine the winding number as the curve goes around the origin once. Since the winding number is 1 due to the eigenvalue at the origin, and since the curve on which the winding number is computed encircles the region where all unstable eigenvalues are trapped, we have provided a strong numerical argument for the spectral stability of the fronts.

Note that in the script HighLewis.m, we have the lines:

```
c.refine = 'on'
c.tol=0.1;
```

The lines above require that the relative differences on the values of the Evans function between two consecutive points on the curve to be less than 10%. STABLAB will adapt the mesh along the curve until the relative difference is less than 10% or until a certain number of tries is attained.

A way to look for unstable eigenvalues while avoiding the problem of venturing into the essential spectrum is to use a contour of the form illustrated in Figure 4.10. The inner semicircle is used to avoid the origin $\lambda = 0$ at which the Evans function has zero. To do so, we use the script semicirc2.m, which was introduced earlier in this section in the case of the Burgers equation. The following lines of codes define the contour illustrated in Figure 4.10 with radius $r = 1$ and inner radius $inner\_radius = 0.01$:

```
circpnts=100; imagpnts=100; innerpnts = 50; r=1;
spread=4; inner_radius=10^(-2);
preimage=semicirc2(circpnts,imagpnts,innerpnts,c.
    ksteps,r,spread,inner_radius,c.lambda_steps);
```

Remember that in the case $\beta = 1$, the bound is found to be 0.9615. This means that any eigenvalue with a positive real part would be trapped in a semicircle region as the one illustrated in figure 4.6 with radius $A = 0.9615$. The winding number is zero (since we are avoiding the origin) as illustrated in Figure 4.15 showing the graph of the Evans function.

As reported in [33, Section 6.2], for the spectral problem related to the high Lewis number combustion model, there are two eigenvalues that travel from the left to the right of the complex plane when β reaches the value 6.572. More precisely, two eigenvalues of the operators (4.28) appear on the imaginary axis at $\lambda = \pm 1.623 \times 10^{-3}i$ when $\beta = 6572$ and start to travel to the right as the value of β increases. The eigenvalues appear as a pair of complex conjugates because of the symmetry of the Evans function with respect to the real axis following from Theorem 4.5. Note that the values of λ at which the eigenvalues cross the imaginary axis are different from the one listed in [33] due to different scaling on the variable t being used. As the values of β continue to increase, the two eigenvalues eventually turn around and start to move left until they go back to the left side of the complex plane [29, Figure 2]. Here, we use the value $\beta = 6.8$ at which the eigenvalues are known to be on the right side of the complex plane. We compute the bound on the norm $|\lambda|$ of any unstable eigenvalue using our program BoundComputation.m. We find a bound of 0.0203. We use the contour defined by:

```
circpnts=100; imagpnts=100; r=0.1; spread=2;
    zerodist=0;
```

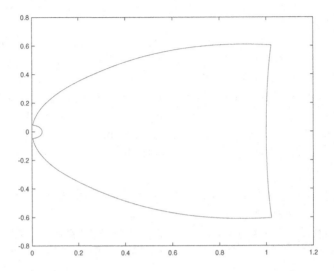

Figure 4.15 Figure obtained when running the script HighLewis.m in the case $\beta = 1$. It depicts the graph of the Evans function associated to the traveling wave solution of the high Lewis number combustion model (2.42) as λ varies along contour illustrated in Figure 4.10 with radius $r = 1$ and inner radius $inner\_radius = 0.01$. The winding number is found to be 0 as the graph does not circle the origin.

```
preimage=semicirc(circpnts,imagpnts,c.ksteps,r,
    spread,zerodist,c.lambda_steps)-0.0001;
```

The contour defined above passes to the left of the imaginary axis in such a way for the conditions (4.96) to be satisfied. In view of the bound of 0.0203, the semicircle as defined above must include any unstable eigenvalue in its interior. We find a winding number equal to 3, which is consistent with the fact that inside the contour, there are three eigenvalues: the one at the origin and the two extra ones that traveled from the left side of the complex plane. We use the following lines:

```
circpnts=100; imagpnts=100; innerpnts = 50; r
    =0.1;
spread=4; inner_radius=10^(-5);
preimage=semicirc2(circpnts,imagpnts,innerpnts,c.
    ksteps,r,spread,inner_radius,c.lambda_steps);
```

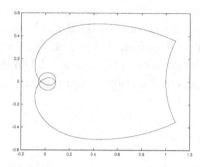 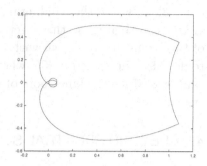

Figure 4.16 Graphs of the Evans function obtained when running the script HighLewis.m in the case $\beta = 6.8$ as λ varies along a contour that includes the origin (on the left) and does not include the origin (on the right). The winding number is found to be 3 for the graph on the left and 2 for the one on the right.

to describe a contour that avoids the origin. The winding number is calculated to be two. This is consistent with the results of [33] since we are excluding the eigenvalue at the origin. Figure 4.16 shows the graph of the Evans function in both cases.

In order to confirm the precise location of the eigenvalue when it crosses the imaginary axis, we use the following line to define a contour about $1.623 \times 10^{-3}i$

```
points=100;
preimage = 0.001623*1i+0.0001*exp(2*pi*1i*
    linspace(0,1,points+(points-1)*c.ksteps));
```

Since the contour is entirely on the upper half of the complex plane, we comment the line that extends the contour for values of the complex plane under the complex by adding to it its reflection with respect to the real axis

```
%w = [w fliplr(conj(w))]; % We compute the Evans
    function on half of contour then reflect
```

The resulting winding number when running the script HighLewis.m is found to be 1, thus confirming the result of [33]. One might be interested in finding the path of that eigenvalue as β increases from $\beta = 6.572$. To do so, one can, for different values of β, trap the eigenvalue inside a circle along which the winding number of the Evans function is equal to 1. One can then make that circle smaller and smaller until the location of the eigenvalue is known with enough precision. A more efficient way to do this is described in [55] where the authors explain how STABLAB can be used for tracking roots of the Evans function as parameters in a system vary.

4.4 BEYOND SPECTRAL STABILITY

Let say you worked through the methods from this chapter and figured out the location of the unstable discrete spectrum. There are three possible options to consider.

If there is an eigenvalue in the open right-half plane of the complex plane (an eigenvalue with a positive real part), then the generic conclusion is that the wave is unstable. Perturbations to the wave are expected to grow exponentially and most likely the wave is not going to survive. It may cease to exist or may be replaced by some other coherent structure, a different wave or a more complex solution. In applications, the unstable waves are rarely observed. Even in laboratory conditions some perturbations are present, since nothing is ideal. Exponential growth is fast, so perturbations soon overtake the unstable wave. From this point of view, stable waves are the ones that are usually observed in real life. In some applications the stable, observable waves are important to know about, for other applications the instability of the wave has strong physical implications so finding out if the wave is unstable is equally important. For example, in population dynamics an unstable wave may not carry realistic meaning since it is not observable, on the other hand in combustion a stability may be either advantageous, for example, when burning of a fuse or for controlled burning, or disadvantageous because a stable wave may be more difficult to extinguish. If the wave is unstable, it is important to understand what this instability means in the context of the application.

Say we found neither essential spectrum nor an eigenvalue with a nonnegative real part with the exception of a simple eigenvalue at zero, so the wave is spectrally stable. Before making a conclusion about the nonlinearly exponentially stability with asymptotic phase as defined in Section 4.1, one should check if the spectral stability in the particular case implies the nonlinear stability, which means that the perturbations to the wave die out not only in the linear approximation of the system but also as solutions of the nonlinear equation. Let us give a precise definition of the nonlinear stability with asymptotic phase.

Definition 4.6 *[87, Definition 7.1] A traveling wave $u = u_0(\xi)$ is non-linearly stable if, for every $\epsilon > 0$, there is $\delta > 0$ with the following property: if $\overline{u} = \overline{u}(x)$ is an initial condition in some neighborhood $\mathcal{E}_\delta(u_0)$, then the associated solution $u(\cdot, t)$ satisfies $u(\cdot, t) \in \mathcal{E}_\epsilon(u_0(\cdot + \tau); \tau \in \mathbb{R})$ for $t > 0$, where neighborhoods $\mathcal{E}_\delta(u_0)$ and $\mathcal{E}_\epsilon(u_0(\cdot + \tau); \tau \in \mathbb{R})$ are possibly measured in different, not necessarily equivalent norms. We say that u_0 is nonlinearly stable with asymptotic phase if, for each $\overline{u} = \overline{u}(x)$ as above, a τ^* exists such that $u(x, t) \to u_0(\cdot + \tau^*)$ as $t \to \infty$.*

In effect, Definition 4.6 says that a traveling wave is nonlinearly stable if any solution with initial conditions close to the wave remains close to it at all times. The term "closeness" is determined by the neighborhoods $\mathcal{E}_\delta(u_0)$ and $\mathcal{E}_\epsilon(u_0(\cdot + \tau); \tau \in \mathbb{R})$. The precise definition of these neighborhoods depends the function space used in the problem. For example, how close the solution is to the wave may be measured in the $L^2(\mathbb{R})$ norm defined in (4.16). However, depending on the problem, other function spaces might be used to determine closeness. In $\mathcal{E}_\epsilon(u_0(\cdot + \tau); \tau \in \mathbb{R})$, the τ is used to allow the solution to stay close to a translation of the wave, $u = u_0(\xi + \tau)$ different from the original wave if $\tau \neq 0$. The last sentence of the definition says what asymptotical stability means: not only does the solution stay close to the wave, but it approaches it as $t \to \infty$.

Spectral stability implies the nonlinear stability with an asymptotic phase if the nonlinearity in the original equation is a smooth function of the unknown functions and the operator obtained by linearizing the equation about the wave belongs to a special class of operators called sectorial [44, Section 5.1]. Without going into details, an operator is called sectorial if its spectrum belongs to a sector of the complex plane with the vertex on the real axis and opens to the left of the imaginary axis and if there are specific bounds on the norm of its resolvent which is an operator associated with the eigenvalue problem. These conditions are not easy to check and require a background in, at least, functional analysis, but there is good news also. For the equation or system of the form $u_t = du_{xx}$, where $d > 0$, the operator defined by the right hand side is known to be sectorial [44]. It is also known that a sum of a sectorial operator and a bounded operator linear in the first-order derivative, as well as linear in the unknown functions produces another sectorial operator. In the case when the original equation is a reaction-diffusion equation

$$u_t = du_{xx} + f(u),$$

the linearization about the wave $u = u_0$ looks like

$$v_t = dv_{\xi\xi} + cv_\xi + f'(u_0)v, \quad \xi = x - ct,$$

and so the right hand side defines a sectorial operator as long as $f'(u_0)$ is bounded and smooth as a function of the variable ξ (recall that $u_0 = u_0(\xi)$ and therefore $f'(u_0)$ is a function of ξ).

A similar result holds for systems of reaction diffusion systems

$$\mathbf{u}_t = D\mathbf{u}_{xx} + \mathbf{f}(\mathbf{u}), \tag{4.107}$$

where D be a diagonal matrix with strictly positive entries, i.e. system (4.107) is parabolic. Let the equation

$$\mathbf{v}_t = D\mathbf{v}_{\xi\xi} + c\mathbf{v}_\xi + J[\mathbf{f}](\mathbf{u_0})(\mathbf{v}),$$

where $J[\mathbf{f}]$ is the Jacobian of $\mathbf{f}$, be the linearization of system (4.107) about the wave $u = u_0(\xi)$. Then as long as the entries in $J[\mathbf{f}]$ are bounded, smooth functions of ξ, the corresponding operator is sectorial.

The following theorem relates the spectral stability of a wave to the nonlinear stability with an asymptotic phase.

Theorem 4.6 *[44, Section 5.1]. If, in some space E, f defines a C^1 mapping, the operator $\mathcal{L}$ is sectorial, and the traveling wave u_0 is spectrally stable, then u_0 is nonlinearly exponentially stable with asymptotic phase.*

Theorem 4.6 in effect says that for a certain class of operators of linearization, the spectral stability (as defined in Definition 4.5) implies nonlinear stability. The assumptions on f are satisfied, for example, if f is a polynomial function of u, because such a function is in C^1.

An example satisfying the conditions of Theorem 4.6 is the Nagumo equation (1.41) and its front solution in the case $0 < a < 1/2$ with boundary conditions given in (4.38). The solution is spectrally stable since the essential spectrum is bounded away from the imaginary axis in the left side of the complex plane (see Figure 4.4). Furthermore the point spectrum is restricted to the set $\text{Re}(\lambda) < 0$, except for an eigenvalue at $\lambda = 0$ as a consequence of Lemma 4.7 of [103], written in terms of the

Evans function (see also [25]). The front thus is spectrally stable and Theorem 4.6 implies its nonlinear stability with an asymptotic phase.

We also point out that the spectral stability of fronts in Nagumo equations may also be deduced based on their monotonicity. The spectral stability of monotone solutions in scalar, bistable equations follows from the Sturm-Liouville theory (for detailed information see Section 2.3.3 of [61]). Sturm-Liouville Theory connects the number of eigenvalues to the right of a given eigenvalue with the number of zeros of the eigenfunction that corresponds to that eigenvalue. We know that the derivative of the front is an eigenfunction that corresponds to the zero eigenvalue because of the translation invariance. If the front is monotone, then its derivative has a constant sign and does not have any zeros, therefore the zero eigenvalue happens to be the right-most eigenvalue. This argument implies that a monotone front in a Nagumo equation has no unstable discrete spectrum.

The same result can be achieved through a different technique. For monotone solutions of scalar bistable equations a general nonlinear stability result exists [25, 98] that follows from the Comparison Principle in the theory of partial differential equations. The result implies nonlinear stability with an asymptotic phase against sufficiently small in supremum norm perturbations for any monotone front in a bistable equation.

For the situation when the system is partly parabolic, i.e. when some but not all of the entries in the diagonal matrix D are zero, the situation is more complicated. Because some of the elements of D in (4.107) are zero, then the essential spectrum of the wave has branches with vertical asymptotes and therefore the spectrum cannot be confined in any sector opening to the left half-plane of the complex plane. The operator is not sectorial and therefore Theorem 4.6 is not applicable.

There is another theorem that implies nonlinear stability with asymptotic phase [6]. The assumptions in this theorem are not easy to check. The operator of the linearization is not assumed to be sectorial, but a different additional condition is imposed on the operator, that involves some heavy functional analysis machinery. In [6] the authors show that the theorem works when the reaction-diffusion equation is partly parabolic, but the traveling wave is a pulse. Fronts need a different treatment. The nonlinear stability result described below in Theorem 4.7 follows from [35, Theorem 3.1] and the result from [6].

The theorem states that spectral stability guarantees that the assumptions from [6] on the linearization operator about of fronts and pulses hold in partly parabolic systems of the form:

$$\partial_t \mathbf{u_1} = D\partial_{xx}\mathbf{u_1} + \tilde{A}\partial_x\mathbf{u_1} + \mathbf{f_1}(\mathbf{u_1}, \mathbf{u_2}),$$
$$\partial_t \mathbf{u_2} = \mathbf{f_2}(\mathbf{u_1}\mathbf{u_2}),$$
(4.108)

where $\mathbf{u_1} \in \mathbb{R}^k$, $1 \leq k < n$, and $\mathbf{u_2} \in \mathbb{R}^{n-k}$, D and $\tilde{A}$ are constant matrices, D is a diagonal matrix with strictly positive entries, and f_1 and f_2 are continuously differentiable.

The linearization of system (4.108) about a traveling wave that moves with velocity $c \neq 0$ is

$$\partial_t \mathbf{v_1} = D\partial_{\xi\xi}\mathbf{v_1} + \tilde{A}\partial_\xi\mathbf{v_1} + c\partial_\xi\mathbf{v_1} + B_{11}(\xi)\mathbf{v_1} + B_{12}(\xi)\mathbf{v_2},$$
$$\partial_t \mathbf{v_2} = c\partial_\xi\mathbf{v_2} + B_{21}(\xi)\mathbf{v_1} + B_{22}(\xi)\mathbf{v_2}.$$
(4.109)

Assume that each $B_{ij}(\xi)$ exponentially approaches a constant matrix $B_{ij}^{\pm}$ as $\xi \to \pm\infty$. This is not a restrictive assumption for the traveling waves that approach their rest states at exponential rates. The linear operator $\mathcal{L}$ associated with the linearization (4.109) is given by

$$\mathcal{L} = \begin{pmatrix} D\partial_{\xi\xi} + \tilde{A}\partial_\xi + c\partial_\xi + B_{11} & B_{12} \\ B_{21} & c\partial_\xi + B_{22} \end{pmatrix}.$$
(4.110)

We have the following theorem:

Theorem 4.7 *Suppose the spectrum of $\mathcal{L}$ from (4.110) is contained in* $\mathrm{Re}\lambda \leq -\nu$, *for some* $\nu > 0$, *except for an eigenvalue 0 of finite algebraic multiplicity. Then the traveling wave is nonlinearly exponentially stable with asymptotic phase.*

The model for burning of solid fuels (2.40) described in Section 2.2.2 satisfies the conditions of Theorem 4.7 in some parameter regimes and when the spectrum is calculated in a space with an exponential weight [36]. Exponential weights are used to move the essential spectrum that touches the imaginary axis to the left of it. Exponential weights create additional difficulties because in an exponentially weighted norm the nonlinear terms in (2.40) lose their smoothness, but nevertheless some additional arguments lead the stability investigation to Theorem 4.7 which implies the nonlinear stability in this case.

Bibliography

[1] M.J. Ablowitz and A. Zeppetella. Explicit solution of Fisher's equation for a special wave speed. *Bull. Math.Biol.*, 41:835–840, 1979.

[2] M.A. Aziz-Alaoui and M.D. Okiye. Boundedness and global stability for a predator-prey model with modified Leslie-Gower and Holling-type II schemes. *Appl. Math. Lett.*, 16, 2003.

[3] B. Barker, J. Humpherys, G. Lyng, and J. Lytle. Evans function computation for the stability of travelling waves. *Philos. Trans. R. Soc. A*, 376(2117):20170184, 2018.

[4] B. Barker, J. Humpherys, J. Lytle, and K. Zumbrun. STABLAB: A MATLAB-based numerical library for Evans function computation. *Available at https://github.com/nonlinear-waves/stablab_matlab*, 2015.

[5] H. Bateman. Some recent researches on the motion of fluids. *Mon. Weather Rev.*, 43:163–170, 1915.

[6] P.W. Bates and C. Jones. Invariant manifolds for semilinear partial differential equations. *Dynamics Reported*, 2:1–38, 1989.

[7] G.O. Batzli, H.-J.G. Jung, and G. Guntenspergen. Nutritional ecology of microtine rodents: linear foraging-rate curves for brown lemmings. *Oikos*, 37:112–116, 1981.

[8] I. Beardmore and R. Beardmore. The global structure of a spatial model of infectious disease. *Proc. R. Soc. Lond. A*, 459:1427–1448, 2003.

[9] P. Belousov. A periodic reaction and its mechanism. In R. Field and M. Burger, editors, *Oscillations and traveling waves in chemical systems*, 605–613. New York: Wiley, 1985.

[10] A.J. Bernoff, M. Culshaw-Maurer, R.A. Everett, M.E. Hohn, W.C. Strickland, and J. Weinburd. Agent-based and continuous models of hopper bands for the Australian plague locust: How resource consumption mediates pulse formation and geometry. *PLOS Comput. Biol.*, 16:e1007820, 2020.

[11] J.P. Boyd. *Chebyshev and Fourier spectral methods*. Courier Corporation, 2001.

[12] F. Brauer, C. Castillo-Chavez, and Z. Feng. *Mathematical models in epidemiology*. Springer, 2019.

[13] R.J. Briggs. *Electron-stream interaction with plasmas*. The MIT Press, Cambridge, MA, 1964.

[14] J.M. Burgers. A mathematical model illustrating the theory of turbulence. *Adv. Appl. Mech.*, 1:171–199, 1948.

[15] G.A. Carpenter. A geometric approach to singular perturbation problems with applications to nerve impulse equations. *J. Diff. Equ.*, 23:335–367, 1977.

[16] P. Carter and A. Scheel. Wave train selection by invasion fronts in the FitzHugh-Nagumo equation. *Nonlinearity*, 31:5536–5572, 2018.

[17] C.C. Conley. On traveling wave solutions of nonlinear diffusion equations. In *Lecture Notes in Phys.*, 38:498–510. Springer-Verlag, New-York, 1975.

[18] E. Díaz, J.M. Amado, J. Montero, M.J. Tobar, and A. Yáñez. Comparative study of Co-based alloys in repairing low Cr-Mo steel components by laser cladding. *Phys. Procedia*, 39:368–375, 2012.

[19] P.G. Drazin, and R.S. Johnson. *Solitons: an introduction*. Cambridge university press, 1989.

[20] A. Ducrot, Z. Liu, and P. Magal. Large speed traveling waves for the Rosenzweig-MacArthur predator-prey model with spatial diffusion. *Phys. D*, 415:132730, 2021.

[21] S.R. Dunbar. Traveling wave solutions of diffusive Lotka-Volterra Equations: a heteroclinic connection in R^4. *Trans. Amer. Math. Soc.*, 286:557–594, 1984.

[22] R.J. Field, E. Kórós, and R.M. Noyes. Thorough analysis of temporal oscillations in the Ce-BrO3 – malonic acid system. *J. Am. Chem. Soc.*, 94:8649–8664, 1972.

[23] R.J. Field and R.M. Noyes. Oscillations in chemical systems. IV. Limit cycle behavior in a model of a real chemical reaction. *J. Chem. Phys.*, 60:1877–1884, 1974.

[24] P.C. Fife. The bistable nonlinear diffusion equation: basic theory and some applications. *Appl. Nonlinear Anal.*, Proc. Third Internat. Conf., Univ. Texas, Arlington, Tex., 1978:143–160, 1979.

[25] P.C. Fife and J.B. McLeod. The approach of solutions of nonlinear diffusion equations to travelling wave solutions. *Bull. Amer. Math. Soc.*, 81:523–532, 1975.

[26] R.A. Fisher. The wave of advance of advantageous genes. *Ann. Eugenics*, 7:353–369, 1937.

[27] R. Fitzhugh. Impulses and physiological states in theoretical models of nerve membranes. *Biophys. J.*, 1:445–466, 1961.

[28] R.A. Gardner and K. Zumbrun. The gap lemma and geometric criteria for instability of viscous shock profiles. *Comm. Pure Appl. Math.*, 51:797–855, 1998.

[29] A. Ghazaryan, J. Humpherys, and J. Lytle. Spectral behavior of combustion fronts with high exothermicity. *SIAM J. Appl. Math.*, 73(1):422–437, 2013.

[30] A. Ghazaryan, S. Lafortune, and C. Linhart. Flame propagation in a porous medium. *Phys. D*, 413:132653, 2020.

[31] A. Ghazaryan, S. Lafortune, and V. Manukian. Stability of front solutions in a model for a surfactant driven flow on an inclined plane. *Phys. D*, 307:1–13, 2015.

[32] A. Ghazaryan, S. Lafortune, and V. Manukian. Spectral analysis of fronts in a marangoni-driven thin liquid film flow down a slope. *SIAM J. Appl. Math.*, 80:95–118, 2020.

[33] A. Ghazaryan, S. Lafortune, and P. McLarnan. Stability analysis for combustion fronts traveling in hydraulically resistant porous media. *SIAM J. Appl. Math.*, 75:1225–1244, 2015.

[34] A. Ghazaryan, S. Lafortune, and P. McLarnan. Combustion waves in hydraulically resistant porous media in a special parameter regime. *Phys. D*, 332:23–33, 2016.

[35] A. Ghazaryan, Y. Latushkin, and S. Schecter. Stability of traveling waves for degenerate systems of reaction diffusion equations. *Indiana Univ. Math. J*, 60:443–472, 2011.

[36] A. Ghazaryan, Y. Latushkin, S. Schecter, and A.J. De Souza. Stability of gasless combustion fronts in one-dimensional solids. *Arch. Ration. Mech. Anal.*, 198:981–1030, 2010.

[37] A. Ghazaryan, Y. Latushkin, and X. Yang. Stability of a planar front in a class of reaction-diffusion systems. *SIAM J. Math. Anal.*, 50:5569–5615, 2018.

[38] A. Ghazaryan, V. Manukian, and S. Schecter. Travelling waves in the Holling-Tanner model with weak diffusion. *Proc. R. Soc. A*, 471:20150045, 2015.

[39] R. Granit. Award ceremony speech, *NobelPrize.org. Nobel Prize Outreach AB 2022*. 1963. Accessed: 2021-07-10.

[40] P. Gray and S.K. Scott. *Chemical oscillations and instabilities: non-linear chemical kinetics, Intern. Ser. Monogr. Chem. 21*, Clarendon Press, Oxford, 1990.

[41] J. Guckenheimer and P. Holmes. *Nonlinear oscillations, dynamical systems, and bifurcations of vector fields*, volume 42 of *Appl. Math. Sci.* Springer-Verlag, Berlin, 1st ed. 1983. Corr. 5th printing, 1997.

[42] G.H. Hardy, J.E. Littlewood, G. Pólya. *Inequalities.* Cambridge university press, 1952.

[43] P.S. Hastings. On the existence of homoclinic and periodic orbits for the FitzHugh-Nagumo equations. *Q. J. Math.*, 27(1):123–134, 1976.

[44] D. Henry. *Geometric theory of semilinear parabolic equations, Lecture Notes in Math., 840.* Springer-Verlag, 1981.

[45] A. Hodgkin. The ionic basis of nervous conduction. *Nobel Lecture. NobelPrize.org. Nobel Prize Outreach AB*, 2021. Accessed: 2021-07-10.

[46] A.L. Hodgkin and A.F. Huxley. Action potentials recorded from inside a nerve fibre. *Nature*, 144:710–711, 1939.

[47] A.L. Hodgkin and A.F. Huxley. Resting and action potentials in single nerve fibres. *J. Physiol.*, 104:176–195, 1945.

[48] A.L. Hodgkin and A.F. Huxley. A quantitative description of membrane current and its application to conduction and excitation in nerve. *J Physiol.*, 117:500–544, 1952.

[49] A.L. Hodgkin and A.F. Huxley. Currents carried by sodium and potassium ions through the membrane of the giant axon of Loligo. *J Physiol.*, 116:449–472, 1952.

[50] A.L. Hodgkin and A.F. Huxley. Propagation of electrical signals along giant nerve fibres. *Proc. R. Soc. Lond. B. Biol. Sci.*, 140:177–183, 1952.

[51] A.L. Hodgkin and A.F. Huxley. The dual effect of membrane potential on sodium conductance in the giant axon of Loligo. *J Physiol.*, 116:497–506, 1952.

[52] A.L. Hodgkin, A.F. Huxley, and B. Katz. Measurement of current-voltage relations in the membrane of the giant axon of Loligo. *J Physiol.*, 116:424–448, 1952.

[53] C.S. Holling. The components of predation as revealed by a study of small-mammal predation of the European pine sawfly. *Can. Entomol.*, 91:293–320, 1959.

[54] J. Humpherys. *Spectral energy methods and the stability of shock waves*. Doctoral thesis. Indiana University, 2002.

[55] J. Humpherys and J. Lytle. Root following in Evans function computation. *SIAM J. Numer. Anal.*, 53:2329–2346, 2015.

[56] R.S. Johnson. A non-linear equation incorporating damping and dispersion. *J. Fluid Mech.*, 42:49–60, 1970.

[57] C. Jones. Stability of the travelling wave solution of the Fitzhugh-Nagumo system. *Trans. Amer. Math. Soc.*, 286:431–469, 1984.

[58] C. Jones. Geometric singular perturbation theory. In *Dynamical Systems (Montecatini Terme)*, 1609:44–118. Springer, Berlin, 1995.

[59] T. Kapitula. Multidimensional stability of planar travelling waves. *Trans. Amer. Math. Soc.*, 349:257–269, 1997.

[60] T. Kapitula. Stability analysis of pulses via the Evans function: dissipative systems. In *Dissipative Solitons, Lecture Notes in Phys*, 661: 407–427. Springer-Verlag, New York, 2005.

[61] T. Kapitula and K. Promislow. *Spectral and dynamical stability of nonlinear waves*, Applied Mathematical Sciences, 457. Springer, 2013.

[62] T. Kapitula and B. Sandstede. Stability of bright solitary-wave solutions to perturbed nonlinear Schrödinger equations. *Phys. D*, 124:58–103, 1998.

[63] T. Kato. *Perturbation theory for linear operators*, volume 132. Springer Science & Business Media, 2013.

[64] B. Kazmierczak and V. Volpert. Travelling waves in partially degenerate reaction-diffusion systems. *Math. Model. Nat. Phenom.*, 2:106–125, 2007.

[65] A. Kolmogorov, I. Petrovskii, and N. Piskunov. A study of the diffusion equation with increase in the amount of substance, and its application to a biological problem. *Selected Works of A. N. Kolmogorov I*, 1:248–270, 1991.

[66] D. Korteweg and G. de Vries. On the change of form of long waves advancing in a rectangular canal, and on a new type of long stationary waves. *Philos. Mag.*, 39:422–443, 1895.

[67] M.D. Korzukhin and A.M. Zhabotinsky. Mathematical modeling of chemical and ecological auto-oscillating systems (In Russian). In *Mol. Biophys*. Nauka, Moscow, 1965.

[68] M. Krupa, B. Sandstede, and P. Szmolyan. Fast and slow waves in the FitzHugh-Nagumo equation. *J. Diff. Equ.*, 133:49–97, 1997.

[69] C. Kuehn. *Multiple time scale dynamics*. Springer, New York, 2015.

[70] S. Lafortune and J. Lega. Instability of local deformations of an elastic rod. *Phys. D*, 182:103–124, 2003.

[71] E. Logak. Mathematical analysis of a condensed phase combustion model without ignition temperature. *Nonlinear Anal.*, 28:1–38, 1997.

[72] M.M. López-Flores, D. Marchesin, V. Matos, and S. Schecter. *Differential equation models in epidemiology.* Collóq. Bras. Mat. IMPA, Brazil, 2022.

[73] A. Lotka. Undamped oscillations derived from the law of mass action. *J. Am. Chem. Soc.*, 42:1595–1599, 1920.

[74] S. Luckhaus and L. Triolo. The continuum reaction-diffusion limit of a stochastic cellular growth model. *Rend. Mat. Acc. Lincei*, 15:215–223, 2004.

[75] W. Mafliet and W. Hereman. The tanh method I - exact solutions of nonlinear evolution wave equations. *Phys. Scr.*, 54:563–568, 1996.

[76] W. Mafliet and W. Hereman. The tanh method II- exact solutions of nonlinear evolution wave equations. *Phys. Scr.*, 54:569–575, 1996.

[77] J.D. Meiss. Differential dynamical systems. *Mathematical Modeling and Computation*, 22. SIAM, 2017.

[78] J.D. Murray. On traveling wave solutions in a model for Belousov-Zhabotinskii reaction. *J. Theoret. Biol.*, 56:329–353, 1976.

[79] J.D. Murray. Mathematical Biology. Springer, 1989.

[80] J. Nagumo, S. Arimoto, and S. Yoshizawa. An active pulse transmission line simulating nerve axon. In *Proceedings of the IRE*, 50:2061–2070, 1962.

[81] J. Nagumo, S. Yoshizawa, and S. Arimoto. Bistable transmission lines. *IEEE Trans. Circuit Theory*, 12:400–412, 1965.

[82] S. Nii. Stability of travelling multiple-front (multiple-back) wave solutions of the FitzHugh–Nagumo equations. *SIAM J. Math. Anal.*, 28:1094–1112, 1997.

[83] L. Perko. Rotated vector fields. *J. Diff. Equ.*, 1:127–145, 1993.

[84] L. Perko. *Differential equations and dynamical systems. Texts Appl. Math.*, 7. Springer-Verlag, Berlin, 2001.

[85] W.H. Press, S.A. Teukolsky, W.T. Vetterling, and B.P. Flannery. Section 18.1. The shooting method. In *Numerical recipes: the art of scientific computing.* Cambridge Univ. Press, 2007.

[86] I. Prigogine and R. Lefever. Symmetry breaking instabilities in dissipative systems II. *J. Chem. Phys.*, 48:1695–1700, 1968.

[87] B. Sandstede. Stability of travelling waves. In: *Handbook of Dynamical Systems II (Edited by B. Fiedler)*, 2:983–1055. Elsevier, 2002.

[88] B. Sandstede and A. Scheel. Absolute and convective instabilities of waves on unbounded and large bounded domains. *Phys. D*, 145:233–277, 2000.

[89] K. Schmüdgen. *Unbounded self-adjoint operators on Hilbert space.* Springer Science & Business Media, 265, 2012.

[90] C.J. Schwiening. A brief historical perspective: Hodgkin and Huxley. *J. Physiol.*, 590:2571–2575, 2012.

[91] L.A. Segel and L. Edelstein-Keshet. *A primer on mathematical models in biology.* SIAM, 2013.

[92] G. Sell and Y. You. *Dynamics of evolutionary equations*, volume 143. Springer, 2002.

[93] S.H. Strogatz. *Nonlinear dynamics and chaos: with applications to physics, biology, chemistry, and engineering.* CRC Press, 2015.

[94] N. Suzuki, M. Hirata, and S. Kondo. Traveling stripes on the skin of a mutant mouse. *PNAS*, 17:9680–9685, 2003.

[95] P. Szmolyan. Transversal heteroclinic and homoclinic orbits in singular perturbation problems. *J. Diff. Equ.*, 92:252–281, 1991.

[96] J.J. Tyson and P.C. Fife. Target patterns in a realistic model of the Belousov-Zhabotinskii reaction. *J. Chem. Phys.*, 73:2224–2237, 1980.

[97] F. Varas and J.M. Vega. Linear stability of a plane front in solid combustion at large heat of reaction. *SIAM J. Appl. Math.*, 62:1810–1822, 2002.

[98] V. Volpert and S. Petrovskii. Reaction–diffusion waves in biology. *Phys. Life Rev.*, 6(4):267–310, 2009.

[99] V. Volterra. Variazioni e fluttuazioni del numero d'individui in specie animali conviventi. *Ann. Eugenics*, 2:31–113, 1926.

[100] R.H. Wang, Q.X. Liu, G.Q. Sun, Z. Jin, and J. van de Koppel. Nonlinear dynamic and pattern bifurcations in a model for spatial patterns in young mussel beds. *J. R. Soc. Interface*, 6:708–715, 2008.

[101] R.N. Wiedenmann and R.J. O'Neil. Laboratory measurement of the functional response of Podisus maculiventris (Say) (Heteroptera: Pentatomidae). *Environ. Entomol.*, 20:610–614, 1991.

[102] A.T. Winfree. The prehistory of the Belousov-Zhabotinsky oscillator. *J. Chem. Educ.*, 61:661–663, 1984.

[103] E. Yanagida. Stability of travelling front solutions of the Fitzhugh-Nagumo equations. *Math. Comput. Model.*, 12(3):289–301, 1989.

[104] W.H. Young. On classes of summable functions and their Fourier series. *Proc. R. Soc. A*, 87:225–229, 1912.

[105] N.J. Zabusky and M.D. Kruskal. Interaction of solitons in a collisionless plasma and the recurrence of initial states. *Phys. Rev. Lett.*, 15:240–243, 1965.

[106] K. Zumbrun. A local greedy algorithm and higher-order extensions for global numerical continuation of analytically varying subspaces. *Quart. Appl. Math.*, 68:557–561, 2010.

Index

$L^2(\mathbb{R})$, 91–93, 95, 97, 103, 111, 112, 145

Absolute instability, 4, 84, 85
Asymptotic phase, 84, 85, 144–148

Belousov-Zhabotinsky reaction, 38, 39, 75
Bistable equation, 20, 147
Boussinesq equation, 132–134
Brusselator model, 39, 75
Burgers equation, 21, 128, 129, 132, 141
Burgers equation, inviscid, 22
Burgers equation, viscous, 22

Closed orbit, 31
Convective instability, 4, 84, 85

Eigenvalue problem, 87, 93
Essential spectrum, 85, 86, 97

Fisher-KPP equation, 14, 17, 26, 40
Fisher-KPP equation, generalized, 17
FitzHugh-Nagumo model, 33

Heteroclinic orbit, 31, 69, 70
Hodgkin-Huxley model, 18
Homoclinic orbit, 31, 69
Hopf bifurcation, 79, 80

Integrable equations, 23

Korteweg-de Vries equation, 22, 25, 89, 90, 94, 95, 98, 100, 104, 105, 126
Korteweg-de Vries equation, generalized, 26

Liapunov number, 79–81
Linear stability, 90, 92
Linearization, 84–86, 88, 90, 94–97, 99, 101, 103, 105–109, 114, 115, 120, 123, 126, 127, 129, 135, 136, 146–148
Logistic equation, 16
Lotka-Volterra model, 34

Nagumo equation, 18–20, 26, 64, 95, 105, 107, 114, 123, 124, 146, 147
Nagumo equation, generalized, 20
Nonlinear stability, 84–86, 144–148

Orbital stability, 3, 83
Oregonator model, 39

Partly parabolic system, 29–31, 33, 46, 147, 148
Periodic orbit, 79
Planar wave, 8, 13
Point spectrum, 97, 98, 102, 103, 107, 119, 120, 146

Quasilinear partial differential equations, 7

Resolvent set, 97

Sectorial operator, 85, 86, 145–147
Semilinear partial differential
    equations, 7
Shock wave, 22
Solitary wave, 22
Soliton, 23, 25, 26
Spectrum, 84, 97
Square-integrable, 91

STABLAB, 128, 129, 131, 132,
    134, 135, 139, 140, 144
Standing wave, 11

Translational invariance, 3, 25, 83,
    94, 136, 140, 147

Winding number, 122, 123, 132,
    134, 137, 139–144

Printed in the United States
by Baker & Taylor Publisher Services